Matthew Cobb is currently a Senior  Sciences at the University of Manch Director for Zoology. Before returning Paris for eighteen years, and has translated several books on biology, evolution from French to English. He writes regularly for the *Times Literary Supplement* and the *Los Angeles Times*. He lives in Manchester with his partner and their two children. This is his first book.

www.egg-and-sperm.com

Further praise for *The Egg & Sperm Race*:

'Anyone wanting to know how the early scientists discovered the secrets of reproduction could not find a better introduction than *The Egg and Sperm Race*'
Sunday Telegraph

'Cobb's exuberant story is welcome ... Wonderfully engaging ... Cobb's lively stories make the process of scientific discovery and adjudication approachable and intriguing'
Nature

'Lively ... You can almost smell the formaldehyde on the page'
Financial Times

'Vivid ... [Cobb] peoples this history with all manner of wonderful, dedicated enthusiasts'
Guardian

'The discovery of sperm and ova and the controversy they generated represent one of the greatest stories in the history of biology. It is a story as relevant today as it was in its own time, and Matthew Cobb tells it with great scholarship and tremendous panache'
Tim Birkhead, author of *Promiscuity* and *The Red Canary*

THE EGG & SPERM RACE

The Seventeenth-Century Scientists
Who Unravelled the Secrets of
Sex, Life and Growth

MATTHEW COBB

POCKET
BOOKS
LONDON • SYDNEY • NEW YORK • TORONTO

First published in Great Britain by The Free Press in 2006
This edition published by Pocket Books in 2007
An imprint of Simon & Schuster UK Ltd
A CBS COMPANY

1 3 5 7 9 10 8 6 4 2

Simon & Schuster UK Ltd
Africa House
64–78 Kingsway
London WC2B 6AH

www.simonsays.co.uk

Simon & Schuster Australia
Sydney

A CIP catalogue record for this book is available
from the British Library.

ISBN 13: 978-1-4165-2600-1
ISBN 10: 1-4165-2600-5

Typeset in Bembo by M Rules
Printed and bound in Great Britain by
Cox & Wyman Ltd, Reading, Berks

For Tina, who won the race twice

CONTENTS

ACKNOWLEDGEMENTS

The people who deserve my greatest thanks are the seventeenth-century thinkers whose work this book describes. Some of these men have been living in my head for nearly seven years, others have forced their way in during the eighteen-month passage from initial idea to completed manuscript. I have tried to hear the voices behind the printed page, to understand their motivations, to sense the personality that drove the hand that scratched ideas on to paper all those centuries ago, and above all to imagine what each one of them did after he put down his pen. Sometimes, the lack of a portrait or of anything more than a few scraps of biographical information made this incredibly difficult. One trick I adopted to force the characters into mental focus was to imagine which actor would best portray each of them (for what it's worth my final list was: De Graaf – Tom Cruise; Harvey – William Hartnell; Leeuwenhoek – Warren Clarke; Redi – Cary Grant; Steno – Alan Rickman; Swammerdam – Nick Cage; Thévenot – Pete Postlethwaite). It has been a privilege to walk with them through their science, retracing their steps towards understanding.

This book would not exist without the perpetual enthusiasm of my agent, Peter Tallack, who was immensely helpful during the long months of writing the proposal, and the insight of my UK editor,

THE EGG & SPERM RACE

Andrew Gordon, who nudged me towards the idea of writing a book about generation. I am incredibly grateful to both of them. My wife, Tina, and my two children put up with me disappearing into the seventeenth century for long periods, while my mother, brother and sister all gave me the support I needed. Allen Moore, then of the Faculty of Life Sciences (FLS) at the University of Manchester, seemed to believe my assurances that the book would not get in the way of my day job (studying the sense of smell in maggots – really), while the members of the Centre for the History of Science, Technology and Medicine in FLS generously encouraged an amateur. Colleagues and friends from my years in Paris, in particular Jean Gayon, Michel Morange and Emma Bayamack-Tam, were supportive of my work in its earliest phases, as were two people I have never met – Clara Pinto-Correia of Lisbon and Edward G. Ruestow of Boulder, Colorado.

My thanks go to librarians in Paris, London, Manchester, Leiden and Göttingen. In particular, Christine Woollet at the Royal Society of London, the staff at the Bibliothèque Inter-Universitaire de Médecine in Paris and at the Special Collection of the John Rylands University Library of Manchester have all been enormously helpful. Perhaps my biggest thanks go to the pioneering spirits at the Rylands Library who decided, long before it was widespread, to subscribe to a massive range of electronic databases. As a result, I could read seventeenth-century texts or search for obscure references from home, in the middle of the night, at the click of a mouse. Reading on-screen or from the print-out of a PDF lacks the delicious contact with history provided by centuries-old paper and can never replace the magic of original manuscripts, but it can be so much more convenient. Joanne McNamara, Ana Paunovic, James Thorne and Marco Zito all translated passages from Latin, making up for my own ignorance. Many other people provided help by answering my questions, in particular Catherine Blackledge, Nicholas Dew, Paula Findlen, Marian Fournier, Cynthia Pyle, Benjamin Roberts and Jane Stevenson. My one regret is that, despite repeated attempts, I could not make contact with Estelle Cohen, whose interests overlap much of this book. The Bibliographical Society generously gave me a grant to study Swammerdam's manuscripts in Leiden in December 2004, where I was able to resolve a number of questions a previous visit had left unanswered, and to feel a kind of communion with Swammerdam's strange soul by walking the cobbled streets of his Leiden, handling his letters and drawings, and tracing his academic career in the university's neat ledgers.

I had the great luck to see some of the episodes in this book brought to life. The Bristol-based Full Beam Visual Theatre's production *The Man Who Discovered That Woman Have Eggs* – the work of Lizzie Philps, Adam Fuller and Rachel McNally – gave me the real privilege of watching the fragmented impressions provided by history turn into living, and hilarious, theatre. Roger Short was incredibly generous in providing me with a copy of his excellent film of Harvey's dissection of a pregnant hind, and allowing me to put a downloadable copy of it on www.egg-and-sperm.com where it will, I hope, attract the worldwide audience it deserves.

Andrew Gordon and my US editor, Gillian Blake, both helped streamline my ideas, reorganising chapters and suggesting painless and stylish cuts. My thanks also go to Edwina Barstow and the designer and proofreader, who made the final stages of production go so smoothly. Roger Short, Cathy McCrohan, Tim Fairs, James Ladyman and Christina Purcell all read through the manuscript and pointed out howlers, both grammatical and factual. My thanks to all of them – I still blush about what might have happened without their help. However, the errors that remain are entirely the responsibility of my darling daughters, Lauren and Evie.

DRAMATIS PERSONAE

Gerard Blaes (1625–92) Head of the Amsterdam Athenaeum school, where he taught Steno and apparently tried to claim credit for one of Steno's discoveries. Persuaded Swammerdam to contribute a drawing of the uterus and ovaries to his textbook, partially precipitating Swammerdam's conflict with De Graaf over the discovery of the human egg.

Reinier de Graaf (1641–73) Catholic physician, trained in Leiden. Brilliant and good-looking, he proved the existence of the human egg in 1672. Had three polemical squabbles over his claims to priority in discovery, first over the structure of the human testicle, then with Van Horne and with Swammerdam over the human egg.

George Ent (1604–89) English surgeon who in 1648 convinced Harvey to publish his study of generation in animals. Later became a member of the Royal Society.

Nehemiah Grew (1641–1712) English physician and botanist who became Secretary of the Royal Society and corresponded closely with Leeuwenhoek over his discovery of spermatozoa.

William Harvey (1578–1657) English physician who studied in Padua, Harvey put forward the hypothesis of the circulation of the blood in 1628. Closely linked to the English court, Harvey published his final work, on the generation of animals, in 1651.

Robert Hooke (1635–1703) 'London's Leonardo', Hooke was a close friend of Christopher Wren and worked with Sir Robert Boyle in the early years of the Royal Society. Published the first popular work of microscopy, *Micrographia*, which was extremely influential.

Christiaan Huygens (1629–95) Dutch physicist and astronomer. Moved to France and joined the Thévenot *académie*, before becoming a leading member of the Académie Royale des Sciences. The first person to bring Leeuwenhoek's discovery of spermatozoa to public attention.

Theodor Kerckring (1640–93) Leiden-trained physician who published an influential but deeply flawed study of human development. Produced little of consequence, but later became a Fellow of the Royal Society and was a close friend of Steno in his final years.

Athanasius Kircher (1601–80) Jesuit priest and polymath who published on a huge range of subjects. His gullible accounts of spontaneous generation helped provoke Redi's study of the generation of insects.

Antoni Leeuwenhoek (1632–1723) Plain-spoken Delft draper who became one of the most productive correspondents of the Royal Society. Using single-lens microscopes he discovered micro-organisms, including bacteria, and in 1677 observed spermatozoa in his own semen.

Marcello Malpighi (1628–94) Italian physician and anatomist who carried out audacious experiments on artificial insemination to prove the roles of semen and eggs. He also made dissections of the developing chick to understand how a fully developed organism can emerge from the egg.

Henry Oldenburg (c.1615–77) German-born Secretary of the Royal Society of London, polyglot and indefatigable letter writer. He encouraged Swammerdam and De Graaf's work, then suggested Leeuwenhoek study semen.

Francesco Redi (1626–97) Physician and poet at the Tuscan court, Redi settled two thousand years of debate over the spontaneous generation of insects and showed the power of the new science.

Niels Steno (1638–86) Dane who made major discoveries in anatomy and natural history, as well as laying the foundations of geology and palaeontology. Trained in Leiden, Steno was Swammerdam's intimate friend and was the first person to suggest that all female animals have ovaries.

Jan Swammerdam (1637–80) Mystical son of an Amsterdam apothecary, he was inspired by Thévenot to study generation. He was particularly interested in insects, which he studied using dissection and microscopes. Trained in Leiden, where he worked with Van Horne on the uterus and ovaries.

Franciscus Sylvius (1614–72) Professor of Medicine at Leiden. Much-loved teacher, he had a chemical theory of disease, allegedly invented gin and was very close to De Graaf.

Melchisedec Thévenot (c.1620–92) One-time French Ambassador to Genoa, polyglot bibliophile, inventor of the spirit level, ex-spy and author of a popular book on how to swim. Prompted Swammerdam and Steno to study generation.

Johannes van Horne (1621–70) Professor of Anatomy at Leiden whose textbooks gave him international fame. Dissected the human uterus and ovary with Swammerdam, and was apparently the first to suggest that the ovaries contained eggs, but failed to publish.

THE EGG &
SPERM RACE

PROLOGUE

You probably don't want to think about this, but, unless you were conceived as a 'test-tube baby', your parents made love about nine months before you were born. Your conception involved a series of events which, until the final years of the twentieth century, were shared by all humans. Somewhere in the world, conception is happening right now.

A few hours ago, a liquid-filled structure in one of the woman's two ovaries popped. It released a minute cell – her egg, and genetically half of the future child. The egg, which was smaller than the head of a pin, had been sitting in her ovary for decades. It was formed, along with thousands of others, before the woman was even born.

Moving from the ovary to the welcoming fronds of the Fallopian tube, the tiny egg was wafted down the narrow tube towards the woman's uterus. This had happened every month since she reached puberty, but this time will be special – the egg will be fertilised and a child will eventually be born.

Perhaps thirty minutes ago, the man ejaculated into the woman's vagina. Ninety per cent of the semen he produced was a mixture of fluids that protect and transport the decisive part of the ejaculate –

tens of millions of minute, wriggling spermatozoa, three months in the making, each one representing a potential other half of the child's genes.

Placed in the upper reaches of the vagina, many of the microscopic spermatozoa are damaged, immobile or in some other way non-functional. Only the strongest have been able to penetrate the thick mucus around the woman's cervix, and to enter the uterus. From here an even smaller proportion have been helped up the Fallopian tube by waves of contractions, until a tiny fraction of the original number of sperm have met the egg. The sperm are now wriggling round the giant egg – each one is only 1 per cent of the size of their female counterpart. They are desperately seeking an entry, blindly attracted to chemical signals given off by the egg.

Suddenly, in a magical process that we still do not fully understand, the egg fuses with one of the sperm. This is not a brutal act of penetration by an aggressive sperm, but a kind of mutual recognition between the two essential components of human life. The head of the sperm disappears inside the egg and the DNA of the two cells – egg and sperm – mingle. Something that can eventually turn into a child has just been created.

However you were conceived, that fusion of egg and sperm all those years ago was the first amazing step in what became your life. What happened next – to you, to me, to all of us who have ever lived – followed more or less the same path. This is how you were made:

In the days following conception, the fertilised egg (or zygote) divided, slowly growing, until after four days there were about fifty cells in a tiny ball.

Gradually, the ball became a small sphere – looking vaguely like a minute egg – and was welcomed into the soft flesh of the uterus wall, which had been building up for this very event, triggered by hormones from your mother's body.

At twenty-one days, the embryo's heart had begun to beat; by eight weeks old, the reproductive organs had formed, and the bones had begun to solidify.

Over the following weeks, the embryo became a foetus, taking on clearly human features; by four months it could move and blink. Soon it could even survive if its birth occurred prematurely.

Most likely, however, you waited until the full nine months were up before moving down towards the exit from your mother's uterus, at which point, on a hormonal signal from her brain, the process of birth began.

You came into the world.

That astonishing story seems to have been inevitable. For good or bad, you were born as you, to your parents. It appears that things could not have been any other way. Except that they could. Your parents' child could have been born the other sex; all it would have taken would be for a sperm of the other kind to fuse with the egg, or oocyte.

And then the fertilised oocyte might not have implanted – this happens to around one-third of zygotes, which simply disappear with the could-be mother's next menstrual period. Even after implantation, nothing is certain; miscarriages (the loss of an implanted embryo) can take place in early pregnancy for many reasons.

And then, even after the embryo overcame all those potential dangers, its experience in the womb could have been different – you could have been a twin (or not), your mother might have smoked (or not), or drunk alcohol (or not), or taken prescription or recreational drugs (or not), or eaten a different ethnic food, or spoken another language.

All of these things, and many others, can affect the experience of the foetus, altering the future child in ways that are both predictable and as yet unknown. What appears with hindsight to have been inevitable was in fact shaped in many of its contours by a series of chances – contingent events which did not have to happen that way, but which did, and thereby contributed to your final form, behaviour, stature and perhaps even taste. The underlying and necessary thread that led from then to now was twisted, shaped and stretched by chance happenings to create a unique human being.

The history of science is like your history. Looking back it can appear linear, smooth, progressive, inevitable. And in one sense it is – you are you, we are where we are. But in another sense, when we look at the fine detail, understanding the role of chance, of errors and of misunderstandings, things are not so straightforward.

When we study history, just as when we study nature, things become more beautiful as we appreciate that what happened was not inevitable. Our sense of wonder grows as we understand how contingent factors are superimposed on a fundamental process. Our vision of events becomes richer, more succulent.

Science did not have to look the way it does. We did not have to arrive at our present knowledge the way we did, just as your parents' child did not have to be you. Understanding how we got here, distinguishing chance from necessity, revelling in the difference and interplay between the two, is exciting and liberating. It shows the role of human choice, initiative and error, all of which modulate the underlying forces that drive events.

This book tells the story of how we discovered the decisive components of the complex chain of reproduction – the existence of spermatozoa and of the human egg and the key processes involved in the development of the embryo. It also shows how scientists realised that human conception is merely one example of the fundamental processes involved in the reproduction and development of all forms of life.

To understand the excitement and commitment of the people who discovered these facts, to appreciate the reasons underlying their motivations and the strength of the emotions that drove them forward, we have to go back to a time when nothing was known for certain about where life comes from. Confusion about the processes involved in sex and growth was so great that conception, reproduction and embryonic development were all collapsed into one unknown and mysterious phenomenon: 'generation'.

Words meant different things back then – simple terms like 'egg', 'seed' or 'semen' have changed their meanings, or have lost their imprecision so that it is now difficult to appreciate the ambiguity that reigned in the past. The challenge of this story is to imagine ourselves in a world that began to disappear 340 years ago, to put ourselves back in a state of ignorance that was dispelled by the work carried out by a network of thinkers in the Netherlands, Italy, England and France.

ourselves in a world that began to develop. Self-awareness is the "instinctive bias in a game of informatics that was disturbed by the work period not by a network of timeless in this Netherlandish Dali Machado and Ham."

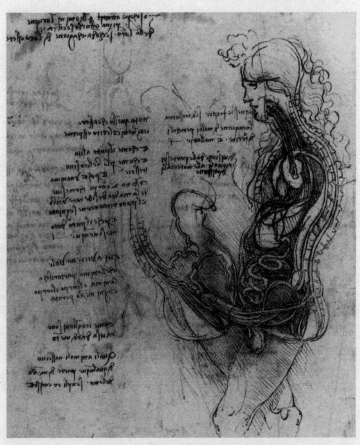

Leonardo da Vinci's drawing of copulation.

1

IN THE BEGINNING

In the Royal Gallery of Windsor Castle there is a drawing by Leonardo da Vinci which shows a cutaway view of a man and a woman having sex. Audacious and stylish, it demonstrates that as well as thinking about helicopters, fluid mechanics and codes, the great Renaissance man was also interested in the fundamental problem of how we come to be. At the top of the drawing, which was made in the 1490s, Leonardo wrote in his famous 'mirror writing': 'I expose to men the origin of their first, and perhaps second, reason for existing.' He was right: sex provides the spark of life and, for many of us, continues to be a major motivating force throughout our adult lives.

But while Leonardo apparently understood the psychology of sex, he was much weaker when it came to the anatomy: not only did he draw the penis straight (in fact it gets bent during intercourse[1]), he showed a non-existent vessel linking the woman's nipples and her uterus, and suggested the man's semen comes from the brain via the spinal cord, instead of being made in the testicles. At the time, confusion about sex was not limited to anatomical errors – throughout history the greatest minds had tried to understand where babies came from, but for Leonardo, as for everyone else, the exact relationship between male, female and offspring was still unclear.

Although people all over the world knew that in general 'like breeds like', nobody knew why, and everyone was prepared to accept that there might be exceptions. In Europe it was generally thought that barnacle geese hatched from barnacles or grew on trees, that insects arose spontaneously from dirt, and that women could sometimes give birth to animals or strange monsters.[2] In the first half of the seventeenth century, the Flemish alchemist Jean-Baptiste van Helmont (1579–1644) gave a famous recipe for generating mice which involved putting a dirty shirt, together with some grains of wheat, into a jar. After twenty-one days, claimed Van Helmont, 'a ferment being drawn from the shirt, and changed by the odour of the grain, the Wheat itself being incrusted in its own skin, transchangeth into mice'.[3] Strange ideas about where life came from, therefore, were not limited to the ill-educated, superstitious masses – they were also held by those at the very top of the intellectual and social pyramid. In the 1660s the Royal Society of London – the world's leading scientific body – discussed how to generate vipers from dust, and was fascinated by the story of the monstrous offspring of a cat and a rabbit.[4]

In this world where like did not necessarily breed like, and where mice could be generated from wheat, there was no space for the idea of 'reproduction'. The word literally means the copying of an individual through the process of generation, and was not used in its current sense until the second half of the eighteenth century. Even trying to investigate the question of generation seemed hopeless; if insects could just appear from nowhere, then it would be pointless to look for any overall logic or pattern. Stuff happened, and that was an end to the matter.

Although the real situation now appears obvious, discovering exactly what goes on was a long, complicated process. Even what might seem to be the most obvious step in 'generation' – the link between sexual intercourse and pregnancy in humans – is really quite difficult to demonstrate. Part of the problem is that clear signs of pregnancy do not immediately follow the sexual act. Even menstruation does not necessarily appear to be directly linked to pregnancy: although women stop menstruating when they are pregnant, some women always have irregular periods, while teenage girls can get pregnant without ever having menstruated. The link between sex and generation is so unobvious that in the twentieth century the Trobriand Islanders in the Pacific Ocean were said to be very surprised to learn that there is a connection between the two.[5] All around the world, folk tales tell of conception taking place in the most astonishing ways, such as by eating fruit (mango, lemon, apple, orange, peach . . .), accidentally swallowing crane dung, or, more poetically, being touched by the rays of a dragon.[6]

It seems most likely that the link between copulation and pregnancy was firmly established by observing domesticated animals when humanity invented agriculture around nine thousand years ago. Animals provide a far better test of ideas about generation than observations of everyday human sexual experience – mating in animals can be controlled by separating the sexes, or can be prevented by castration. Above all, in domestic animals mating always coincides with ovulation, so there is a good chance that copulation will lead to fertilisation. Slowly, over thousands of years, but with growing certainty, mating and generation – at least in domesticated animals and humans – became linked in humanity's collective knowledge.

Once this was understood it was a logical step to assume that the male ejaculate – the only clearly and immediately observable product of copulation – played an important part. This was certainly the view of the biblical writers of the Book of Genesis, who explained how Onan 'spilt his seed on the ground' in order to avoid making his sister-in-law pregnant – the word 'seed' was originally used to refer to both plant seeds and animal ejaculate ('semen' means 'seed'). In one piece of Ancient Egyptian mythology, male semen was literally a seed; after failing to catch the goddess Isis, the god Set ejaculated on the ground, upon which his semen grew and turned into a plant.[7] This widespread identification of the male ejaculate with plant seed was understandable but completely mistaken. Plant seeds are not 'semen' – they are the product of sexual reproduction, rather than merely one component of it. However, the semen/seed analogy proved incredibly powerful and dominated all subsequent thinking about generation. Even today, when young children ask where babies come from embarrassed parents may say that Daddy plants a 'seed' in Mummy's tummy.

But when people thought about it, identifying semen with seed did not really help; if male semen was 'seed', how could children look like their mother? Other questions about generation soon followed. Why did women stop menstruating when they were pregnant, while female birds produced eggs even when they had not mated? What was menstruation? And although everyone knew that birds come from eggs, how did the yellow and white stuff in an egg turn into a baby bird? All these issues were a matter of dispute and speculation for thousands of years. Imagine yourself trying to answer these questions from scratch, using simple observations of the natural world, and you get some idea of the scale of the problem facing humanity. This shows the importance of the leap forward in our knowledge that took place in the second half of the seventeenth century, which created the context that enabled us to solve the puzzle of generation.

Seventeenth-century physicians and philosophers were not the only people at the time to be intrigued by the question of generation. Writers were equally fascinated. In the 1630s, the English poet Thomas Randolph wrote a poem in praise of his friend and patron, Ben Jonson, in which he listed the great mysteries of life, placing generation up there with the transmutation of metals and the motion of the stars:

> How Elements does change; What is the cause
> Of Generation; what the Rule, and Laws
> The Orbs does move by;[8]

Around fifteen years later, Izaak Walton in his classic *The Compleat Angler* revealed the scale of contemporary confusion when he suggested that pike, eels and carp are generated from reeds.[9]

More importantly, the question of generation was of crucial importance for women. Childbirth was dangerous, and by the seventeenth century there was a growing European market for books, pamphlets and broadsheets about childbirth and other aspects of women's health.[10] Unfortunately, most of the material relating to generation was misleading and out of date. In 1654 the final edition of Thomas Raynalde's best-seller, *The Birth of Man-kinde; Otherwise Named the Womans Booke*, was published.[11] This book, which was explicitly aimed at women readers, had first appeared more than one hundred years earlier. The genteel ladies who were able to read the book were given useful modern advice about dealing with 'Dandruffe of the head' and 'the ranke savour of the arme-holes', but Raynalde's vision of generation was essentially that of the Ancient Greeks – there was not a modern idea to be found. Things were even worse for peasant and working women; although the overall level of ignorance about health, anatomy and basic physiological processes affected all layers of the population, it was ordinary women who suffered most from the folk remedies and the ineffective medicine of the age.[12]

Within half a century, the situation had changed radically. In 1707, James Drake's medical textbook *Anthropologia Nova; or, A New System of Anatomy* gave a modern account of human generation, attempting to explain the relative contributions of male sperm and female egg, and providing a roughly accurate outline of embryonic development.[13] Between the final edition of Raynalde's book and the first edition of Drake's, an extremely productive period of scientific investigation had laid the bases of our modern understanding of sex and growth. Although traces of these new ideas could be found in popular literature,[14] this new knowledge was generally limited to the European intellectual elite: for centuries to come most of the population continued to be influenced by pre-scientific ideas. Women would have to wait until the nineteenth century before they could consult a physician who had even a vaguely correct idea of how the female body functioned, while it was only in the twentieth that generalised education, informed by scientific and medical developments, enabled women fully to understand conception – and how to avoid it.

The story that follows tells how we uncovered the basic laws of generation in the 1660s and 1670s. Much of the work described here understandably focused on human reproduction, but it was set in the broader framework of the generation of all life, from flies to frogs. By the end of the period covered in this book, the key discoveries that shape how we now view the natural world were established, but their full meaning and implications remained obscure. Part of the paradox of science is that, often, discovery does not immediately lead to understanding; the full impact of a particular finding may become apparent only many years afterwards, and can take even longer to become common currency. The book explores this contradiction, showing how seventeenth-century scientists directly shaped our understanding of life, where it comes from and how it grows, but failed fully to comprehend the truth that was staring them in the face.

Up until the second half of the seventeenth century, European thought on virtually every question relating to the natural world was dominated by the ideas of the Ancient Greeks. In the case of generation there were two conflicting views, each going back to the great flowering of Greek and Roman culture.

In the fifth century BC, Hippocrates, the founder of medicine, argued that generation took place through the joint action of two kinds of semen – one provided by the man (the ejaculate), the other by the woman (her menstrual blood). This accounted for the few facts that were certain about generation (men ejaculate, menstruation stops during pregnancy), although Hippocrates mistakenly thought menstruation was the key female contribution to generation. A century after Hippocrates, Aristotle came up with an ambitious account of generation in all animals, according to which mammals were generated when the male's semen combined with the female menstrual blood to form the new individual, with the heart – the most important organ – being formed in the womb instantly after copulation. As opposed to Hippocrates, Aristotle suggested that the male and female contributions were not equivalents – they were not both 'semen'. In Aristotle's view the female provided the matter that would constitute the foetus, while the male's semen contributed the form, shaping and sculpting the embryo.[15] The apparently decisive objection that offspring sometimes look more like their mother than their father was met with the response that plants looked different depending on the soil they were grown on, thus showing that matter (soil or the menstrual blood) could clearly affect the form in some cases.

Aristotle argued that the generation of 'lower' animals, such as insects, was very different, as they were generated spontaneously from decay. They reproduced in this way, he said, because they were much simpler than large animals, having no internal organs. This corresponded fairly closely to everyday experience – maggots seem to appear suddenly in rotting matter, and insects are apparently devoid of any inner structures. Spontaneous generation also implied that insects did not 'breed true', because they did not actually breed at all – they were generated by chance from decay. According to Aristotle, there was nothing rational about generation in this part of the natural world.[16]

In the second century AD, Galen – a Greek physician living in what is now Turkey – created a new theory of disease and bodily function. Galen's ideas synthesised an approach to diagnosis and treatment that influenced Western science and medicine for the next fifteen hundred years. Based on Hippocrates' suggestion that the human body functioned on the basis of four 'humours' – blood, phlegm, yellow bile and black bile – and that disease resulted from an imbalance of these humours, Galen's system prescribed purges and blood-letting to re-establish the balance.

The central difference between the views of Aristotle and Galen with regard to human generation was the role of the woman – for Galen, like Hippocrates, she contributed something similar to the man, a female 'semen'. But unlike Hippocrates, Galen did not say the menstrual blood was 'semen', but instead claimed women had an internal secretion which was very similar to male semen. One consequence of this was that production of the woman's semen was sometimes argued to be linked to female orgasm, with conception taking place when both partners enjoyed a sexual climax, implying that sex was supposed to be a mutually pleasurable activity.[17] The suggestion that both sexes produce semen was linked to Galen's mistaken view that the male and female genitalia were simply inverse copies of each other. As Galen put it in his instructions for dissecting genitals: 'Turn outward the woman's, turn inward, so to speak, and fold double the man's, and you will find the same in both in every respect.'[18] Yet despite sometimes being so wide of the mark, Galen's ideas, together with those of Aristotle, dominated Western thought for nearly two millennia.

A few hundred years after Galen, the West entered a period of stagnation – the Dark Ages. Although European culture produced lasting works of art, poetry and philosophy during this time, overall the Continent failed to develop the ideas of the Greek and Roman civilisations. Between roughly AD 400 and 1400, the key advances in humanity's understanding of the world occurred elsewhere, in particular in China and in the Arab world. One great exception was the question of generation, on which humanity made little progress. Chinese medicine emphasised complementary bodily processes rather than locating functions in clearly defined bits of anatomy, so in most respects the Chinese understanding of generation was less clear than that of Aristotle and Galen.[19] Arab thinkers made fundamental contributions to mathematics, optics, chemistry and the teaching and practice of medicine, but like their contemporaries in China they were weak when it came to anatomy and its function, and they were relatively uninterested in investigating the natural history of animals and plants.

The Renaissance of European culture that took place from the fourteenth century onwards was marked by a resurgence of the ideas of Aristotle, Galen and other ancient thinkers. For example, Leonardo's suggestion that semen comes from the man's brain can be traced directly back to the ideas of the school of Pythagoras in the sixth century BC, which were kept alive and transmitted via the Arab world. But although the Renaissance gave rise to a new spirit of openness and discovery in all realms of culture, it also saw Aristotle's ideas about the natural world turned into a stifling orthodoxy that crushed initiative and investigation, mostly due to the work of Thomas Aquinas, an Italian monk and philosopher who lived in the thirteenth century. Aquinas argued that the natural world could be understood by using the ideas and method contained in Aristotle's philosophy, while spiritual truths such as the Resurrection could be known only by faith and the teachings of the Church. This view, which became known as the 'Thomist dogma', led the Church to defend and promote all of Aristotle's ideas, including those that had no immediate impact on theological issues. Among the positions that came under the protection of the Church was Aristotle's version of the common-sense impression that the Sun and stars go round the Earth, according to which the heavens rotated above our heads on gigantic 'spheres'. To challenge Aristotle was to challenge part of the Church's authority – 350 years after the death of Aquinas, this was the trap that closed around Galileo when he defended the idea that the Earth goes round the Sun. The Church could have accepted such a Sun-centred vision of the universe without damaging its theology (it eventually did, in 1992), but once it had decided to approve Aristotle's theory, defending its power became more important than defending the truth.

Thomas Aquinas was also interested in generation, particularly with respect to the moment at which the soul entered the human foetus – 'ensoulment'. Aristotle had argued that this happened at around forty days in a boy and about three months in a girl (the female embryo allegedly grew more slowly). Aquinas adopted this view, which was integrated into Christian thinking at the time. The Catholic Church changed its opinion on the matter only in the nineteenth century, when, faced with new scientific discoveries, it decided that the single cell of the fertilised egg constitutes an 'ensouled' human being, thereby setting itself against any form of abortion and today against the therapeutic use of embryonic stem cells.[20]

After Aquinas, thinkers struggled to graft the views of Aristotle and Galen on to the natural world, while maintaining theological purity about the nature of Christ. The Dominicans adopted an Aristotelian position on Christ's origin, suggesting that Mary provided the material that formed Christ's body, while the Holy Spirit shaped and animated him. Some Franciscans, on the other hand, took Galen's position on generation to its logical conclusion and argued that Mary had actually contributed an active 'semen' to the creation of Christ.[21]

Not everyone agreed with this scholastic approach to the world. Theophrastus Philippus Bombastus Aureolus von Hohenheim (1493–1541), better known as Paracelsus, was one of the key figures to argue that a new kind of knowledge was needed. Paracelsus was an alchemist, occult thinker and itinerant physician who criticised what passed for medical training in most European universities – the rote learning of the works of Galen and Aristotle.[22] Instead, Paracelsus adopted ideas that went back to Arab philosophy and claimed that natural objects had 'correspondences' – innate sympathies which governed the relations of all things, and in particular the response of the patient to medication. Rejecting Galen's theory of humours, Paracelsus sought to understand disease and treatment in terms of the interaction between these similar or opposite characteristics.

Like other alchemists, Paracelsus thought that humanity had lost the knowledge that we had when we lived in the Garden of Eden. To recover that knowledge, to understand how each set of 'correspondences' worked, book learning and classical logic were useless. True knowledge could come only from practice, claimed Paracelsus. Understanding the causes of natural phenomena was a religious duty, he argued, which would require the naturalist or physician to enter into a mystical 'union' with the object of study by conducting experiments. This insistence on the importance of experiments was Paracelsus's key contribution to the development of modern science.

When it came to the question of generation, Paracelsus had a recipe for generating a miniature human being, or 'homunculus', which involved allowing human semen to putrefy for four weeks in a warm sealed container. The result would be something that was 'in some degree like a human being, but, nevertheless transparent and without body.'[23] The next step was to feed this thing with human blood for another forty weeks, after which 'it becomes thenceforth a true and living infant, having all the members of a child that is born from a woman, but much smaller'. The idea was to demonstrate the power of alchemy – it could even create life – and to fabricate a creature that could act as a servant, a bit like the golem in the Jewish tradition. In fact, behind the mystical nonsense and the spooky ambition, this was basically an application of Aristotle's view of generation. Ironically, Paracelsus's vision of generation appeared ridiculous once the decisive criteria of seventeenth-century science – experimental and observational testing, the very criteria he encouraged – came to dominate intellectual life.

The word 'science' was originally used to describe any form of knowledge, and took on its current meaning only in the middle of the eighteenth century. 'Scientist' came into use only in 1840. In terms of their ideas and their professional status, none of the people mentioned in this book were 'scientists' in the modern sense of the word (however, for the sake of simplicity they will often be described as such). In the seventeenth century thinkers studied 'natural philosophy' (what we would now call physics, mathematics and chemistry) or 'natural history' (the life sciences). They employed a variety of methods, but increasingly they used something like what we call science – the progressive cycle of hypothesis, experimentation and theoretical explanation – which began to appear in its recognisably modern form in seventeenth-century Europe in a long process which is generally described as the scientific revolution.[24]

This slow change in the way the world was understood flowed directly from the growth of capitalism and its effects on society. New technology, the explosion of knowledge and communication that followed the invention of the printing press, and the voyages of exploration and conquest which opened up vast new areas of the globe and led to the discovery of astonishing new life forms, all contributed to the development of science. But perhaps the most important factor was the challenge to political, economic and religious authority represented by the new merchant class. This class encouraged the rise of Protestantism, strove for political representation and broke free of the antique economic and social restrictions imposed by the feudal system. The challenge that was implicit in the rise of capitalism altered the intellectual horizons of thinkers all over Europe. Virtually none of the participants in the scientific revolution were either proto-capitalists or politically radical – most of them were politically and religiously conservative, and many were members of the existing aristocratic establishment.[25] Nevertheless, it is no coincidence that the main powerhouses of the scientific revolution – England, the Dutch Republic and some of the Italian states – were also the places where the new merchant class first established itself, and where the printing and publishing traditions were strongest. This link was even clearer in England and the Dutch Republic, which both experienced revolutions against the established order.

By the early decades of the seventeenth century, the new way of investigating and understanding the world had gradually begun to emerge. In Italy, Galileo defended Copernicus's hypothesis about the rotation of the Earth round the Sun and joined a small group of thinkers in Tuscany to form the world's first scientific society, the Accademia dei Lincei (the Academy of the Lynxes – the name suggested that their sight was as sharp as the animal's, able to pierce the darkness). The Lincei began to make use of new instruments like the microscope to study the natural world,[26] while in England, the philosopher Sir Francis Bacon outlined how science should function, using experimentation and objective reporting, and William Harvey put forward the hypothesis of the circulation of the blood. These separate, uncoordinated steps laid the basis for the huge advances in scientific knowledge that took place from 1660.

✳

Although science is the most powerful and most influential form of knowledge humanity has yet invented, scientists generally have little understanding of the history of their subject. Because of the cumulative nature of scientific knowledge, they often see past ideas as leading inexorably to present-day knowledge. The history of science as written by scientists tends to focus on the brilliant discoveries of a handful of men who single-handedly forged our present understanding. At one level, this is justifiable – scientific insights are often the product of one or two individuals, and examples of past brilliance can be exciting and inspirational.

But this vision misses out some of the most interesting parts of history, and presents an oversimplified version of science, past and present.[27] Science never follows a direct path. Instead of progressing straight to some new truth, it takes strange detours, finding itself temporarily trapped in unexpected dead ends. It unwittingly restricts itself to addressing only some of the potential questions that can be asked, simply because it either does not see any other questions or has no means of addressing them. Mistakes play a fundamental role in shaping the form of science – when they reveal themselves, they require new theories to explain why they are in fact mistakes. This gap between theory and reality provides the space for knowledge to develop.

Furthermore, the history of science, like history in general, often hinges on chance. This was particularly true for the history of generation. If Dr George Ent had not felt 'weary with anxiety'[28] during Christmas 1648, the discovery of how generation works would have taken a very different turn.

In that cold winter, Ent, aged forty-four and a leading member of the College of Physicians, decided 'to wipe that dark shadow from [his] mind' by travelling 'not far from the City' to visit his old friend William Harvey. This simple social visit would lead to the first step towards the discovery of the scientific basis of generation. At the time Ent came calling to cheer himself up, Harvey was a 70-year-old with long white hair and twinkling dark eyes who suffered badly from gout. Ent and Harvey knew each other

William Harvey (1578–1657), painted around 1650.

well – they had first met in Italy in 1636, and in 1641 Ent had published a book that defended Harvey's view of the circulation of the blood. When the two men met on that December day, Harvey was initially cheerful, but soon became gloomy as he reflected on the state of the country, admitting that without his studies and his memories he did not know 'what could prevail upon me to wish to survive the present time'.[29]

Although Harvey's use of experimentation in his work on the circulation of the blood had helped give him a radical scientific reputation, his personal life was extremely staid and conservative – he was a staunch royalist and had been physician to James I, and then to his son Charles I, with whom he struck up a close friendship.

Because Harvey's wealth and power flowed from the throne, he lost virtually everything when the Parliamentarian forces were victorious in the English revolution. The new rulers of the country branded him a criminal and forbade him from visiting the City of London – he had to abandon virtually all his medical practice and found himself in much reduced circumstances.[30]

Ent tried to cheer his friend up by asking him to talk about his work. Harvey responded enthusiastically: 'It has always been my delight to make strict inspection into animals,' he said, before launching into a monologue on the impact of the discovery of the New World on the development of science. Ent finally interrupted the old man, reminding him that 'many learned men who are acquainted with your tireless industry in the advancement of natural philosophy are eagerly awaiting the results of your further experiments'. Harvey did not immediately rise to the bait, but smiled and pointed out that in the past his scientific work had aroused controversy and that he had come to value his peace and quiet. What Harvey did not do, however, was deny that he had anything more to publish. Sensing this, Ent teased and flattered him some more: 'Yes but to deserve well and receive ill is the usual reward of virtue,' he said, going on to remind Harvey that in the past he had seen off his critics (with Ent's active support). At this point, Harvey's pride must have got the better of him, for he brought out a huge bundle of papers, covered in his notoriously bad handwriting.[31] 'Now I have obtained what I desired!' exclaimed Ent, as Harvey showed him his final piece of work, *Exercitationes de Generatione Animalium* ('Disputations touching on the generation of animals'), the world's first attempt to understand reproduction and development using experiments and clear reasoning rather than prejudice and tradition.

Initially, Harvey told Ent he did not want the manuscript published, as it was incomplete – a section dealing with the tricky problem of generation in insects had been destroyed during the Civil War. As Harvey put it wistfully:

> While I was attending upon our most serene Highness, the King, in
> our late troubles and more than civil wars, some rapacious hands, and
> that not only by Parliament's permission, but by its command, not
> only spoiled me of all my household goods but also, and this is the
> heavier cause of my lament, my enemies stole from my study my
> notes which had cost me many years of toil. And so it came about
> that very many observations which I had made, particularly those
> concerning the generation of insects, were lost, to the prejudice I may
> boldly say of the commonwealth of learning.[32]

However, despite his initial reticence, Harvey – pleased, flattered,
brow-beaten or a combination of all three – eventually gave the
papers to Ent, 'with absolute power either to publish them forthwith
or for a while to suppress them'. Ent, who had clearly received the
Christmas present he had been hoping for, later wrote that he
'returned [Harvey] very many thanks' and then 'took my leave and
departed like a second Jason enriched with the Golden Fleece'.[33]

Ent was right to feel elated. Harvey's book, which it took Ent two
years to publish, even with the author's help,[34] was an immediate
bestseller, with four separate Latin editions appearing in 1651 (two in
England, two in the Netherlands), followed by an English translation
two years later. De Generatione was a long work for the period –
about twice as long as the book in your hands – and at times it was
very difficult to grasp Harvey's point.[35] Divided into seventy-two
chapters, with no illustrations (which did not make it easier to
understand), De Generatione was a theoretical and experimental
examination of animal generation. The book was a typical seventeenth-
century mosaic, containing elements that were modern and others
that were pre-scientific and superstitious. It included anatomical
descriptions, studies of behaviour (including an excellent description
of mating in the ostrich) and, decisively, experimental studies of generation in chickens and in the deer of King Charles I, with
dissections before and after copulation.

Although Harvey fully accepted the intellectual authority of the
Greeks – 'Respect for our predecessors and antiquity at large teaches

Frontispiece to the first edition of Harvey's De Generatione Animalium *showing Zeus freeing all creation from an egg. The right-hand panel shows a close-up, and the inscription 'Ex ovo omnia' – 'All things come from an egg'.*

us to uphold their doctrines in so far as truth will allow', he said[36] – his aim was to find proof that would enable him to abandon Aristotle's 2000-year-old vision of generation, in which the male's semen acted on the female's menstrual blood to form the embryo. As he stated at the outset: 'That these are false and rash assertions will soon appear: and they will instantly vanish like phantoms of the night when the light of anatomical dissection dawns upon them.' Harvey argued that the key to understanding all generation was the existence of eggs (*ova*, in Latin). He wanted to show that in both animals that produce live young ('viviparous' organisms) and in egg-laying ('oviparous') species, generation involved eggs: 'all animals whatsoever, even viviparous creatures, nay, man himself, are all engendered from an egg, the first conceptions of all living creatures from which the foetus arises are some kind of egg,' he wrote. This view was summed up in the frontispiece of the first edition, in which Zeus is shown releasing all kinds of animals from a huge egg. Written on the egg is a phrase that came to be associated with Harvey, even if he did not state his opinion quite so pithily: '*Ex ovo omnia*' – 'All things come from an egg'.[37]

Over a century later, in 1767, Voltaire wrote that as a result of Harvey's work 'it was accepted throughout Europe that we come from an egg'.[38] This was not quite right – although Harvey's work was ground-breaking, and he did use the term 'egg' with regard to human generation, what he meant by 'egg' was very different from what we – or even Voltaire – would mean. For Harvey the human 'egg' was not the female contribution but rather the early foetus, because he thought it looked like an egg without a shell. He therefore argued that 'all animals are generated after the same manner from an egg-like rudiment'.[39] In other words, all animals come from an egg-shaped embryo, which grows as a consequence of something that happens during mating. Harvey was convinced that the 'something' was not the menstrual blood, but he knew that to find out what it was he would have to do an experiment. His emphasis on the power of experiment was the decisive and most influential part of his book, the methodological dynamo that made it a classic. He showed that theories about generation were not enough – facts were needed, derived from experimental studies.[40]

However, when it came to the generation of insects and 'lower' animals, Harvey presented no experimental data and stuck closely to Aristotle's belief in spontaneous generation. Insects might come from 'eggs' or 'worms', Harvey argued, but they 'arise by chance, by Nature's will, and, as they say, by an equivocal generation from parents different from themselves'.[41] In other words, like did not necessarily breed like, at least in insects. His explanation was that insects 'arise and are procreated forth from beginnings that cannot be seen by reason of their smallness, and seeds scattered and blown hither and thither by the winds, like atoms floating in the air'.[42] By saying that insects could be generated spontaneously, Harvey showed that even if the frontispiece of his book declared '*Ex ovo omnia*', he did not have a consistent idea of what an egg was, nor where it came from. This problem appeared even more clearly in the heart of his book, which dealt with generation in mammals.

In the early 1630s, Harvey had studied the herds of red and fallow deer that were kept in the parks of King Charles I. He dissected hinds that had recently mated, and to his surprise found nothing in the uterus: no semen, no menstrual blood, and most surprising of all, no egg. Studies of dogs and rabbits gave the same result. Harvey summed up his finding succinctly: 'Nothing at all can be found in the uterus after copulation for the space of several days.' This led him to conclude that 'the foetus does not arise from either the male or the female sperm emitted in coitus, nor from both of them mixed together, as the physicians think, nor from menstrual blood as being the substance, as Aristotle thought, and that something of the conception is not necessarily made immediately after coitus.'[43] Harvey's failure to find semen or eggs after copulation led him to believe that both Galen's and Aristotle's views were wrong: whatever the male and female contributions might be, there was no physical contact between them. Furthermore, however this indirect effect worked, it did not immediately lead to the appearance of an egg.

Harvey groped with analogies to explain his findings, suggesting that semen had its effect through some immaterial 'spirit' or 'contagion', or perhaps like an odour, or a spark, or a bolt of lightning, or even a kind of magnetic effect. His vague conclusions were perfectly in keeping with his findings (no semen was observable in the uterus) and with contemporary knowledge about the apparently non-material transmission of disease. As to what the woman's contribution might be, Harvey was at a loss. He dismissed Galen's idea that there was a female 'semen' that was produced at orgasm, and in particular he opposed the suggestion that it involved female 'ejaculation': not all women 'ejaculate', he pointed out; those who do not can still both reach orgasm and be fertile; and he argued that the fluid involved had 'a serous and watery consistency, like urine', which he thought meant it was too thin to play the role of semen.[44]

Harvey made a spectacular mistake when he argued that the female 'testicles' (what we call ovaries) played no role in female generation. His dissections of female deer had shown no changes in the size or shape of the 'testicles' as the mating season progressed, so he had to conclude that 'The so-called testicles, like things utterly unconcerned in generation, neither swell up nor vary in any wise from their wonted constitution either before or after coitus, nor give any indication of being of any use either for coitus or for generation.'[45] This fitted in with the widespread assumption that the female 'testicles' were like male nipples – vestigial, functionless organs.

Ironically, on this point Aristotle was more right than Harvey: the Greek philosopher had described how peasant farmers would sterilise their female pigs 'with the view of quenching in them sexual appetites and of stimulating growth in size and fatness'.[46] The procedure – which Aristotle reported could also be done in camels – involved cutting 'the lower belly about the place where the boars have their testicles',[47] then reaching into the female's body and cutting off a small piece of the ovary. Once stitched up, the animal recovered and went about (most of) its business as usual. The gelding of female domesticated animals was widely practised, including in seventeenth-century England. Harvey even discussed the matter with a sow-gelder, but the profound implications of this piece of folk knowledge do not seem to have struck him, or anyone else.

In the 1970s, Professor Roger Short examined Harvey's findings in the light of modern knowledge and produced a remarkable film in which he repeated Harvey's study.[48] Short concluded that it would have been virtually impossible for Harvey to have come to a correct understanding of the nature of generation, because the deer is one of the worst species to start from if you want to understand mammalian reproduction. For example, male deer will start their mating behaviour – the rut – around a month before the female comes into oestrus, ovulates and mates. Studying females during the rut would not have revealed changes in their ovaries. The apparent lack of an 'egg' in the uterus after mating can be explained by the fact that for weeks the deer embryo does not appear at all egg-like; rather than being an egg-shaped blob of cells, it looks more like a thin piece of snot which is easily overlooked amongst the mess of dissection. Finally, without a microscope to detect the minute spermatozoa, it was hardly surprising that Harvey could find no trace of the male's semen in the female's uterus when he did dissect recently mated hinds.

Faced with this evidence – or, rather, lack of it – Harvey found himself driven into a corner. In the final section of his book, he speculated about what he called the 'dark business'[49] of conception. Struck by his inability to find any physical trace of the future embryo in his dissected deer, nor any sign of semen, Harvey grappled with the implications: 'There is nothing which can be perceived by the senses in the uterus after coitus, and yet it is a necessity that something must be there to render the female fertile.'[50] His conclusion, which he clearly felt uncomfortable with, was the only one his evidence supported. 'What imagination and appetite are to the brain,' he wrote, 'the same thing, or at least something analogous to it, is awakened in the uterus by coitus and from this proceeds the generation or procreation of the egg.'[51] In other words, the appearance of the egg was the product of a mysterious force: the power of the womb. Rather than replacing Aristotle's view of generation with a truly scientific vision, Harvey had ended up with something that looked like a non-materialist version of Aristotle's ideas: semen acted in an immaterial way, while eggs were generated like thought. His step forward had led nowhere; he sought facts, but found confusion.

Harvey's study of generation excited and perplexed its readers. It showed the power of the experimental method and placed the question of generation at the heart of scientific debate, but all that power seemed to be used to no very great effect. His inability to come to any clear conclusion about how generation took place was pretty dispiriting to anyone who wanted to study the subject –

if Harvey had not been able to make any decisive progress, then lesser thinkers stood no chance. Nevertheless, fifteen years after the publication of his book, a group of audacious young men took up the challenge.

FRENCH CONNECTIONS

The horse strained against the rope, and the covered barge slowly moved into the still dark waters of the canal, its flags fluttering in the breeze. Inside, the passengers began to sing from the song sheets provided by the barge's owners. It was the end of September 1661, and the leaves on the canal-side trees were beginning to turn yellow and fall into the water. Among the people making the ninety-minute trip from Amsterdam to Leiden was a 24-year-old man. Like millions of young people since, he was leaving home to go to university. And as many hope but few achieve, he was going to make his mark on the world. Not as a clergyman, despite the wishes of his father, nor as a physician, despite going to Leiden to study medicine. His name was Jan Swammerdam (roughly pronounced Zwar-muh-dum), and together with two friends he met at Leiden he would radically change our understanding of generation. The three young men differed in virtually every way – age, background, personality, ambitions: Swammerdam, the mystical son of an Amsterdam apothecary, was fascinated by insects; Niels Stensen (known in English as 'Steno'), was a serious young Dane with a massive intellectual appetite; Reinier de Graaf, the youngest at twenty, was good-looking and brilliant, a Catholic whose interest was focused on medical and therapeutic questions.[1] What united them was a passion for the new experimental science they learnt from their teachers, and their ambition to make an impact on our understanding of the natural world and the workings of the human body.

The fact that the Netherlands played such a central role in uncovering the secrets of generation might seem surprising. Today Holland is often seen as a quiet backwater, renowned for its liberal attitudes to sex and drugs, but making little contribution to world culture. The contrast with the seventeenth century could not be greater. Then the Netherlands was one of the most powerful and influential countries in the world. Its merchant shipping fleet covered the oceans, transporting expensive exotic products and defended by a navy that was second to none. Its factories produced pottery and cloth using revolutionary methods, its publishing industry was the most powerful and prolific in the world, while its political system, born of revolt and revolution, was one of the most progressive yet created. This was the Dutch Golden Age.

At the same time as Dutch economic power swept aside all competitors, Dutch science, philosophy and culture flourished as never before, whilst religious and philosophical tolerance made the Netherlands a magnet for dissidents of all sorts. This was the time of Rembrandt, Vermeer and hundreds of other artists, a period when seventy thousand paintings were painted and sold each year. Some of the greatest masterpieces humanity has ever produced, paintings that speak to us across the centuries, were created in the Dutch Republic during the heady middle decades of the seventeenth century. The radical philosopher René Descartes (1596–1650), who was the first to realise the importance of mathematical and mechanical models to explain life, left his native France for the Netherlands. The Dutch citizen Baruch Spinoza (1632–77) put forward a radical humanist materialism – a manifesto for tolerance and understanding that has rarely been equalled. This was no medieval backwater, nor an early modern version of today's laid-back Amsterdam. It was a country that was running headlong, eyes wide open, into the arms of modernity, conscious and proud of mapping out the route to the future that the rest of the world would have to follow.[2]

At the origin of this incredible cultural energy was the massive growth of Dutch capitalism, based partly on its local industries, but above all on its huge fleet and its trade with the Far East and the New World. This mercantile activity and its transformation into the world's first example of industrial production was the direct fruit of revolution.[3] In 1572 the mainly Protestant Netherlands had rebelled against Spanish control and the religious and financial oppression that went with it, leading to a long armed struggle against the Spanish Empire. With independence assured in the 1630s, the Dutch economy entered a new phase of expansion: state taxes declined as the armed forces were reduced in size, while Dutch maritime domination brought in huge profits through the state-backed trading company, the United East India Company (Verenigde Oost-Indische Compagnie – the VOC). The Netherlands soon had a monopoly on the supply of nutmeg, mace and cloves – key items in a world avid for strong new tastes.

Strange substances from exotic locations became regular features in Amsterdam's ledgers – indigo from Coromandel, cochineal from Mexico, cinnamon from Ceylon – forming an evocative sketch of the international commercial web woven by Dutch capitalism. This network was built upon a very simple principle: never move an empty ship. Every journey had to bring in a profit, no matter how complex the tangle of traffic. As a result, the patterns of trade were positively mind-bending: Dutch vessels carried pottery from Holland to Spain; they sold the pottery for silver which they then took halfway round the world to Japan, where they used the money from the silver to buy copper, which they transported to Ceylon; in Ceylon they offloaded the copper and bought elephants which were then taken to Bengal; in Bengal they sold the elephants and then loaded up with raw silk, Dacca muslin and opium for the return trip to Europe.

There was a darker side to the wealth generated by Dutch trading.

The VOC's western equivalent, the WIC (West-Indische Compagnie – West India Company), also made money from the global transport of goods, but above all the WIC was the largest slave-trading organisation in the world. Although slave-ownership was illegal in the Netherlands, the slave trade was not, and the activities of the WIC enjoyed the substantial military support of the Dutch state. The Dutch used the supply of slaves to service their plantations in South America and the Caribbean, which produced tobacco, coffee and cotton which were in turn brought back to Europe. The forced labour of enslaved populations generated money at every step. All the fruits of the Dutch Golden Age – including the profound humanism of Rembrandt, the limpid beauty of Vermeer's work and the great discoveries of Dutch scientists – were framed by the decimation of African populations and the unimaginable misery caused by the slave trade.

The revolution in political and economic organisation which swept the Dutch Republic to the forefront of the world stage had its counterpart in the development of science. The Protestant belief that the Bible could be studied by everyone, and not simply by priests, had an effect on the study of nature – here, too, everyone was thought to be able to make a contribution to understanding the natural world. Following the invention of the microscope in the Netherlands sometime in the early seventeenth century, many leading members of Dutch society became important amateur contributors to the new science of microscopy – Johann Hudde (1628–1704), who was Burgomaster of Amsterdam for thirty years, was a skilled microscopist and mathematician, while Spinoza was also an expert grinder of microscope lenses.

Throughout Europe at this time, there was a direct link between personal wealth and science, in the shape of 'cabinets of curiosities' – private collections of weird and wonderful objects from around the world.[4] These cabinets were particularly popular in the Dutch Republic, with more than eighty recorded at the height of the Golden Age.[5] One of the most famous examples was that of Jan Swammerdam's father. It occupied an entire floor of the family house in Amsterdam and contained more than three thousand items, from eggs to abortions, from Chinese porcelain to Roman coins, from fossils to puffer-fish, and including a 17-foot snake. Although these collections now seem an exotic mixture of stamp collection and freak show, they were the direct ancestors of museums. By assembling unusual and wonderful objects, such cabinets showed that the world was full of infinite wonders that could be discovered by those with the desire and the means to study them.

The scientific heart of the Dutch Republic was in Leiden. Threaded with wide canals lined with tall trees, crossed by beautiful brick bridges and bordered by broad quays, the city oozed wealth and had a justified reputation as the 'cleanest and nicest town in the whole of Europe', although in summer the discharge of raw sewage into the canals could lead to an overpowering stench.[6] A major source of Leiden's wealth was the textile industry, which employed around twenty thousand people, many of them in that amazing modern invention – the factory. This made Leiden the second-largest manufacturing city in Europe, after Lyon. Women and girls made up 30 per cent of the workforce, earning lower wages than their male comrades. However, the new bosses were keen to show off their progressive credentials – in 1646 they generously limited the working day for children to a mere fourteen hours. Despite the oppression and exploitation, living standards could be higher than in the surrounding countryside: driven by the growth of manufacturing, the population of Leiden grew rapidly, increasing by 20 per cent in twenty-five years to reach 72,000 in 1672.[7]

The wealth produced by the textile workers and consumed by the burghers made Leiden a very attractive place. It also helped pay for the city's university, which was one of the most radical sites of learning on the face of the planet. From its foundation in 1575, Leiden University emphasised both the traditional subjects (theology, philosophy and law) and history, mathematics and medicine, attracting the brightest scholars and embracing the new values of tolerance, humanism and the importance of research that were to strike so many visitors to the Netherlands. From relatively modest beginnings – there were only one hundred students in 1590, crammed into an old priory – the university rapidly expanded, with more than five hundred students per year by the 1640s, mainly studying law and theology. Between 1626 and 1650 11,000 students (all of them men, around half from abroad) passed through the university. As a comparison, only 8400 students went to Cambridge over the same period.[8]

From the outset, the small Leiden medical faculty was highly progressive. It adopted radical theories and teaching methods, and undertook two ambitious construction projects.[9] In 1587 the university authorities agreed to construct a botanical garden for the medical faculty. This was a research tool for studying the medicinal potential of the thousands of exotic plants that were brought back from the Dutch voyages of discovery and exploitation. Physicians needed to be able to identify plants and their potential therapeutic functions; by creating such a large garden, Leiden University was staking its claim to be at the forefront of European medicine. Indeed, the very existence of the botanical garden highlighted the university's ambition to seek out new knowledge.

The other prestigious construction was a purpose-built amphitheatre for teaching human anatomy. The sixteenth century had seen a number of major changes in how anatomy and dissection were taught. At around the same time as Vesalius's pioneering book on human anatomy *De Humani Corporis Fabrica* appeared in 1543, medical professors literally came down to ground level and started getting their hands dirty. Previously, the professor had read out a text by Galen or Aristotle, his knowledge often based on dissections of monkeys and spiced up with superstition and hearsay. Meanwhile, a 'demonstrator' did the bloody work, hacking away at a body, trying desperately to find the structures that the book claimed were there.

The inadequacies of medical education were not resolved by more reliable textbooks or by a hands-on professor: access to knowledge was still restricted by the simple lack of space – only so many people could get round a dissected body to see what was going on inside.

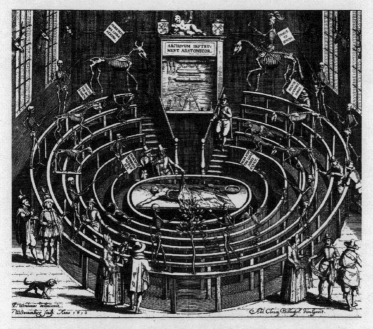

The anatomy amphitheatre, Leiden Medical School. The perspective in this picture is deceptive: the amphitheatre was far narrower and taller than it appears here. The skeletons were arranged like this during the summer, when no dissections could take place because of the heat. The artist, however, has not been able to resist portraying a corpse on the dissecting table. The smart lady at bottom right is showing great interest in a skin that has been flayed from a human being.

The anatomy amphitheatre at Leiden changed all that. Modelled on a similar structure at Padua, the tall, circular construction sat inside the top floor of the old priory. Composed of six rings of steeply banked tiers, with benches on the first two rows and standing room for the rest, the amphitheatre could hold up to two hundred. To keep people's minds on the solemnity of what they witnessed – the dissection of a human body – the room surrounding the amphitheatre was dominated by a human skeleton holding a banner that read 'Birth is the beginning of death.' In the summer months, when the bodies decomposed too quickly for dissection to be practical, the amphitheatre was decorated by several other human and animal skeletons, including a man riding on a horse.

The amphitheatre was not just for the few dozen medical students. The rest of the seats would have been occupied by visiting physicians, town worthies and the ghoulishly curious.[10] In the winter months there could be two or three dissections a week[11] as disease and cold took their toll of the elderly and weak, and the professors wielded their scalpels not only to inform and educate, but also to entertain. This was the world that shaped the scientific outlook of Steno, Swammerdam and De Graaf in the 1660s. They learnt their trade on the benches of the amphitheatre, staring at the body on the black and red turntable which was within arm's reach of the lucky spectators on the front row, looking at the waxy face and the white limbs and torso, made whiter by the pale light falling through the tall windows.

Steno was the first of the three friends to come to Leiden, early in 1661. Born in Copenhagen on New Year's Day 1638, the son of a goldsmith, Steno first went to medical school in his home city, where he became interested in dissection and in the latest chemical theories of disease and treatment. In 1660, having completed three years' training, he followed the path of many other Danish students and travelled to Amsterdam where he studied at the Athenaeum, a medical academy run by the City Physician, Gerard Blaes. Disappointed by the education he received there – he later said Blaes had taught him nothing – Steno soon moved to Leiden, where he enrolled in the medical faculty. During his time with Blaes, he discovered the duct that brings saliva from the parotid gland into the mouth, which was later named 'Steno's duct' by the Leiden professors. The following year, Blaes tried to take the credit for this discovery. Steno defended his claim in a book describing his dissections of glands in the mouth and nose.[12] At a time when so little scientific anatomy was known, even this relatively pedestrian collection had an impact, and Steno was immediately fêted in both the Dutch Republic and

Steno (top left) and De Graaf (top right). There is no known portrait of Swammerdam; above is an imaginary portrait of him by the eighteenth-century Dutch artist Jan Stolker. The face is taken from Rembrandt's painting The Anatomy Lesson of Dr Tulp, *which was painted five years before Swammerdam was born (see Cobb, 2000).*

his native Denmark – his former teacher, the Danish chemist Ole Borch, described him as 'a genius'.[13]

As Steno was making a name for himself, Jan Swammerdam arrived in Leiden. His father initially wanted Jan to be a clergyman, but in the first of many struggles between the two men Jan refused, and was eventually allowed to study medicine. Swammerdam inherited some of his father's collecting mania: he too had a cabinet, but, unlike his father's ragbag of exotica, Jan's collection focused on what was to be his obsession throughout his life: insects.[14]

The youngest of the trio was Reinier de Graaf, born in July 1641 in the town of Schoonhoven, to Catholic parents. Although Dutch tolerance meant that Catholics were free to worship, the Republic was a staunchly Protestant state and, for example, Catholics were not allowed to teach in universities. De Graaf would have grown up with a very distinct impression of being part of an oppressed minority.[15] His father, Cornelis, was a successful naval architect and inventor of hydraulic machines, while his mother, Catharina, came from a wealthy family. After beginning his studies at the Catholic University of Leuven, and having already completed eighteen months' preliminary medical study at Utrecht University, De Graaf arrived in Leiden in February 1663, shortly before Steno left, beginning his medical studies as Swammerdam was finishing his.

Although the three men were clearly friends, they were in different years, with different religious beliefs and different characters, and never formed a tight-knit group. Swammerdam and Steno were very good friends, experimenting on dogs and frogs, and discussing the latest discoveries; Steno presented Swammerdam to Ole Borch when his former teacher visited Leiden in 1662.[16] De Graaf was admired by both staff and students for his skill and intelligence, but he was nonetheless set apart from the two older men.

Student life in 1660s Leiden was remarkably similar to that of students in the twenty-first century, apart from the absence of female students.[17] Young men were crammed into private lodgings, often as many as ten per house, and, like modern students, they tended to change address quite often – Swammerdam lived in at least four different houses in his two years as a Leiden undergraduate. Student life revolved around studying, drinking, dancing late into the night and sex. Virtually any social gathering – playing cards, fencing, playing tennis or going to the theatre – was a pretext for getting drunk. Alcohol consumption even had the official seal of approval: Leiden

A Dutch student's rooms in the middle of the seventeenth century.

University gave students an annual tax-free alcohol allowance of 194 litres of wine and around 1500 litres of beer, as a way of enticing young men to enrol.

In the smoke-filled taverns (smoking tobacco in clay pipes was all the rage) the medical students would be the target of black humour about the failings of their chosen profession – 'We don't have a doctor, we prefer to die free of charge' was a typical joke of the time. Evenings could also be enlivened by the 'cat' game: a hapless feline would be shoved into a cage hung from the rafters; the clients would hurl clubs at the cage until it broke, the cat fell to the floor and was massacred. Bets were placed and a good time was had by all. Except the cat. As well as downing their 'thick beer', the young students would eat pickled fish, cheese, peas, beans, cabbage, beets and turnips. And after the pub, drunken students, like other youths, would fall in the canals and sometimes visit prostitutes. Far more seriously, they would get into fights – in May 1662 a philosophy student was expelled after a night-time brawl in which a local man was killed.

When the 'cat' game palled, or they ran out of money for betting or drinking, the students could flirt in pubs and in the street. In 1663 the English student John Ray visited Leiden and reported that 'the common sort of women (not to say all) seem more fond of and delighted with lascivious and obscene talk than either the English or the French'.[18] This kind of talk, coupled with racy games such as 'Head in the Lap', in which a man placed his head in one girl's lap and then guessed which of a gaggle of other girls was smacking his bottom, led to many goings-on. At all levels of society, sexual attitudes were complex, with marriage generally signalling an end to such licentious behaviour amongst women. As Ray put it: 'The women are said not much to regard chastity whilst unmarried, but when once married none more chaste and true to their husbands.'

Nevertheless, student life was not all play. In 1663 medical students found the timetable hard going. From 8 to 10 a.m. each day Professor Van Horne lectured on therapeutic methods, followed by an hour with Professor Vortius on the aphorisms of Hippocrates. Then came two hours of dissection with Professor Vander Linden, followed (except on Sundays) by a two-hour hospital visit with Professor Sylvius. The day closed with two hours of study with Vortius in the botanical garden. Students were also expected to enrol with each professor for private extracurricular tuition, at their own cost.[19]

In the 1660s, European medical knowledge and teaching were in a state of flux. New anatomical and physiological discoveries such as the circulation of the blood had undermined confidence in the tradition of Galen and Aristotle without necessarily replacing it with anything more coherent or effective. As an English physician lamented in 1648: 'All the order of teaching is troubled and the doctrine of Physick is endeavrd and learned altogether preposterously and confusedly, without any certain method.'[20] Medicine remained in a pre-scientific state, particularly in the decisive realm of treatment. Physicians generally did more harm than good, not only because their 'therapies' were frequently bizarre and at best pointless, but above all because they had no idea what they were doing or how it might affect the patient's health. Many nooks and crannies of the human body were still unexplored, and understanding of the main bodily functions had not progressed for fifteen hundred years. As a result, the more radical universities such as Leiden considered it their task to find modern, mechanical and materialist explanations of respiration, digestion and movement, in order to understand disease.

Leiden was lucky in having on its staff two key representatives of this modern approach to medicine – Johannes van Horne and Franciscus Sylvius.[21] Van Horne had become professor at the young age of twenty-nine, and had written a widely translated introduction to anatomy. He came from an obscenely wealthy family – his father had been one of the sixteen original directors of the Dutch East India Company, and was one of the most powerful men in the Netherlands. Although fascinated by the latest discoveries in science, Van Horne was more interested in anatomical detail than in any overall theoretical or methodological approach.

Probably the best-loved teacher of medicine at Leiden was Sylvius, who taught diagnosis and treatment. Born François dele Boë in Hanau, near Frankfurt, he adopted a Latinised version of his name when he was a student at Leiden medical school ('Sylvius' is based on the Latin word for 'wood', 'Boë' being the way the French word *bois* (wood) was often spelt at the time). Sylvius became renowned for his chemical approach to medicine – he interpreted bodily functions, disease and treatment in terms of chemical reactions, in particular as examples of effervescence produced by the interactions of acids and alkalis. Sadly, despite linking the latest discoveries in anatomy with his radical chemical theory, and frowning on harmful practices such as bleeding, Sylvius was just as ineffective as other physicians. In his inaugural lecture of 1658, he ruefully noted that university-trained physicians were generally no more successful than folk doctors. In Sylvius's case this was hardly surprising: on the basis of his chemical

Professor Sylvius of Leiden University.

theory, he decided that most diseases were due to an excess of alkali; his treatment for both the plague and syphilis, therefore, was to neutralise the alkali in the body by administering vinegar. This had the great virtue of doing no harm (no mean thing in the seventeenth century), but it would have done nothing to halt the progress of the disease.

Sylvius was unusual because he actually studied patients and their symptoms: he took the medical students to St Cecilia's hospice, where he showed them patients and invited them to diagnose illnesses by asking the patients questions. For Sylvius, sick people were not mere objects, but suffering human beings who should be at the heart of the physician's concerns. As well as taking students out to hospitals, Sylvius also showed them how to concoct medicines (it is widely believed that he invented gin, but the evidence is poor). This practical teaching eventually took place in a purpose-built chemistry laboratory which was constructed in the late 1660s after a long battle with the university authorities. While waiting for building to begin, Sylvius had installed a distilling room and a laboratory full of chemical equipment in his spacious new house on the Rapenburg canal (number 31, it still has his coat of arms on the front). As befitted a sophisticated, cultured man, his home was designed around his two great passions, science and art: he had a collection of 140 paintings which had to compete for space with his laboratories.[22]

After two years of studying anatomy, diagnosis and so on, medical students had to write a doctoral thesis, based on a research topic. Steno was exempted from this requirement because of his previous work, and he was awarded his doctorate *in absentia*, in 1664.[23] Swammerdam and De Graaf, however, had to go down the usual route, and chose very different subjects, each on the cutting edge of contemporary medical knowledge. Their choices reveal something of their opposing characters, and their selection of supervisor underlined the differences between them. Swammerdam, studying under Van Horne, chose to investigate breathing, while De Graaf worked with Sylvius on a key organ in his acid–alkali system, the pancreas. Swammerdam's interests lay in pure research, while De Graaf chose a subject with clear therapeutic implications. De Graaf was businesslike and professional, submitting his dissertation in December 1664, shortly after graduation. Swammerdam, meanwhile, took over four years before submitting his dissertation (the final date was February 1667, three and a half years later than expected).

Swammerdam's and De Graaf's studies were similar in one way: they both involved massive invasive surgery on live, unanaesthetised dogs. It is difficult to know what must have been worse: the whining and yelping of the helpless animal, its rolling, terrified eyes, its pathetic attempts to free itself from the ropes that held it down, the blood gushing from the wound, the stench of the faeces and urine that terror squirted from its body, or simply the sheer inhumanity of it all. The people who conducted such experiments were deeply troubled by the horror involved. As Steno put it in a letter to a friend: 'I must admit it is not without abhorrence that I torture them with such prolonged pain. The Cartesians take great pride in the truth of their philosophical system, but I wish they could convince me as thoroughly as they are themselves convinced of the fact that animals have no souls!!'[24] Surgery on a human patient would hardly have been much different (anaesthetics were nearly two hundred years in the future) – except that the patient would have given his or her consent. The dog had no such choice. And, in a way, neither did scientists at the time. Either they went through with these appalling experiments, or science would remain at a standstill. There really was no alternative for understanding the basic processes of bodily functions.

De Graaf's thesis on the pancreas, based on his dog experiments, was a perfect combination of experimental ingenuity and theoretical ambition.[25] When the contents of the stomach move into the intestine, they encounter fluids from the bile duct (coming from the liver and the gall bladder) and from the pancreas. On the basis of his acid–alkali theory of human physiology, Sylvius deduced that because bile is alkaline, the pancreatic juices must be acidic, with the two secretions producing fermentation (frothing) and heat when they meet. De Graaf set out to demonstrate this key component of Sylvius's theory by collecting substantial quantities of pancreatic juice from a live dog. To do this he cut open the dog's belly, shut the pancreatic duct with a small iron clamp, punched lots of holes in the pancreas with a needle, and then tied the gland to a small glass bottle, which hung down outside the dog's body. Once the dog had been sewn up and the wound had healed, the pancreatic juices could easily be collected and studied – a similar bottle was connected to the salivary glands so that both ends of the digestive process could be studied.

Partly because of its experimental audacity, De Graaf's thesis was a big success, appearing in an expanded French edition (1666), a further expanded Latin edition (1671) and an English translation (1676).

But the heart of his reasoning was wrong. He did not demonstrate that the pancreatic juice was acidic, he merely stated it. And his justification was poor: it had to be acidic because this fitted in with Sylvius's theory. In other words, rather than facts driving theory, as the scientific method insisted, theory was driving 'facts'. Worse, the pancreatic juice turns out not to be acidic but slightly alkaline, largely composed of bicarbonate, the rest being enzymes (the key hormone produced by the pancreas – insulin – is secreted directly into the bloodstream and does not pass through the pancreatic duct). The large-scale heat-producing acid–alkali interactions that Sylvius thought were at the heart of physiology were the product of his imagination. On this occasion, De Graaf's fidelity to his teacher's ideas distorted his judgement. De Graaf was undoubtedly brilliant, but he was also fiercely loyal to Sylvius, bound by a complex web of intellectual and emotional links to his teacher.

Swammerdam's work on respiration also involved an experiment on a live dog, in order to study the function of the muscles that expand and contract the chest. With no laboratories in the university, the bloody dissections were carried out in Swammerdam's lodgings – which may explain his regular changes of address during this period. When his thesis finally appeared, four years after the work was carried out, it was an immediate success[26] and subsequently appeared in at least four editions. In the twentieth century it was described as 'one of the classics of the history of physiology'[27] despite the fact that it contains a number of fundamental mistakes (for example, it states that air enters the lungs because the expanding chest compresses the surrounding air, whereas, in reality, the expansion of the chest creates lower pressure within the chest cavity, thus sucking air in).

The most striking part of Swammerdam's thesis, which was also the most important premonition of his future work on generation, was the title page. It is a gothic mixture of vivisected dogs (one of them breathing under water) and strange pieces of glassware. At the bottom of the page are two snails. This illustration, which is a mixture of the accurate and the fantastic, has little to do with respiration. The key point is that they are copulating: this refers to

Title page of Jan Swammerdam's doctoral thesis De Respiratione *('On respiration' – second edition, 1679).*

Swammerdam's brief announcement in the final part of his thesis that he had discovered that some snails are hermaphroditic (the twisted structures are the penises of the two snails; in reality snail copulation is never as tidy as this). By including this discovery at the front of his thesis, Swammerdam was showing that by 1667 his interests had extended far beyond the subject of his doctoral research. He had embarked on a highly ambitious new investigative programme, encompassing the whole of natural history and focusing on the most fundamental mystery that humans have ever encountered: generation. The inspiration for that project had not come from his Leiden professors, but from a visit to France. Coincidentally, Steno and De Graaf made the same journey at the same time, with similar consequences.

Summer 1665. Paris. The rich had escaped the stifling stench of the city heat while the poor were reduced to bathing in the stinking Seine in an attempt to keep cool. More than five hundred thousand people lived crammed into an area about one-fifth the size of the present-day French capital, while on the other side of the city walls there were small villages dotted between fields and vineyards. But despite its rural setting, seventeenth-century Paris was not a pretty place. Outside the gilded palaces and stately homes, the public places were smelly, filthy and drab. Unlike Amsterdam or London, there were no brightly painted shop and inn store fronts to distract the attention from the narrow, unpaved streets, which were ankle-deep in mud and excrement, both animal and human.[28] The streets were not only unpleasant, they were potentially lethal; with no pavements, pedestrians ran the risk of being crushed by carriage wheels or run over by sedan chairs. And when the number of carriages, chairs, carts and horses became too great, the city would seize up in a massive gridlock that could last over an hour.[29] Into this pungent, dense contradiction of crowding poverty and incredible wealth had come Steno, Swammerdam and De Graaf, intent on meeting the members of the scientific discussion circles which had sprung up in the French capital.

In the 1660s the scientific revolution had stepped up a gear, as thinkers began to be interested not only in making discoveries but also in encouraging discovery, by creating scientific societies. London was the centre of these changes, adopting a model that would be admired throughout the world: the Royal Society of London and its journal, the *Philosophical Transactions*.[30] As its Secretary put it in 1664, the Royal Society was 'a Corporation of a number of Ingenious and knowing persons, by ye Name of ye Royall Society of London for improving Naturall knowledge, whose dessein it is, by Observations and Experiments to advance ye Contemplations of Nature to Use and Practice'.[31]

The origins of the Royal Society lay in the tumultuous period of the English revolution and the Commonwealth, when a number of young thinkers – including Sir Robert Boyle, a young Irish noble-man and son of one of the richest men in the country – met in Oxford to carry out experiments. With the restoration of the monarchy in 1660, this 'Invisible College', as they called themselves, sensed an opportunity to gain prestige and, with a bit of luck, money. They approached the new King, Charles II, and asked for his support. However, despite providing the Society with a royal char-ter in the summer of 1662, Charles never came up with the cash. Nor was he really interested in the Royal Society's work: he was more inclined to go to the theatre and laugh at the latest satirical play that ridiculed its occasionally eccentric activities.[32] Charles was pro-foundly mistaken; probably the most long-lasting legacy of his reign was the creation of the Royal Society, which by the mid-1670s had more than two hundred members from England and abroad. Much of the subsequent influence, success and shape of science can ulti-mately be traced back to the Royal Society's work.

Scientific societies influenced the development of science by providing criticism and advice on work in progress and making it possible to carry out experiments that were too complex or expen-sive for one person to do, sometimes commissioning research or the construction of a piece of apparatus.[33] Their meetings and their jour-nals were a way of transmitting findings to other scientists, leading to

the creation of a scientific community. They even forged the informal rules that govern how science functions, emphasising the
importance of explaining how an experiment was done and therefore the ability of others to replicate the finding. By explaining or
even performing an experiment in front of the Society, scientists
could use their peers as a 'collective witness' to gain acceptance for
controversial discoveries.[34]

The creation of the Royal Society had an immediate international
impact. Throughout Europe, thinkers looked to this new, official
organisation, which was able to gather around it some of the greatest minds from England and abroad. Interest was particularly keen in
France. Although there were informal scientific groups in Paris and
in a number of other French cities,[35] generally known as *académies*,
they were more like discussion clubs. Up until 1664, the most
important of these clubs had been the *académie* of Henri-Louis
Hubert de Montmor, which met more or less weekly between 1657
and 1663 in Montmor's house on what is now the rue du Temple in
the centre of Paris. From its very beginnings, however, the Montmor
circle was marked by an uneasy tension between those who sensed
the importance of the new scientific method and wanted to carry
out experiments using scientific instruments, and those who were
content to carry on in the old way, using argument and theory to
settle disputes. By the end of 1663 this difference had erupted into
a series of spiteful rows and the Montmor *académie* had collapsed.

In 1664 and 1665, a series of letters circulated in Paris asking
Louis XIV to create a 'Compagnie des Sciences et des Arts'
(Company of Sciences and Arts). The aim of this grouping would
be 'to do experiments and to discover as many novelties as possible,
in both heaven and earth by Astronomical and Geographic
observations using telescopes, microscopes and all other necessary
instruments'.[36] Within two years, the Finance Minister, Colbert,
had responded positively by creating the Académie des Sciences,
complete with a massive budget from the crown.[37]

The author of the letters is thought to have been Melchisedec
Thévenot, one of those extraordinary, multitalented people who simply

cannot exist in the twenty-first century – patron of the sciences, polyglot traveller, bibliophile, orientalist, one-time French Ambassador to Genoa, inventor of the spirit level, author of a popular book on how to swim, and ex-spy. Thévenot's contemporaries also thought he was extraordinary: the poet Jean Chapelain called him an 'excellent person ... a highly honourable man, highly knowledgeable, who knows all the languages of Europe'; the Dutch astronomer and physicist Christiaan Huygens said he was 'the best and the most honest man one could meet'; while for the philosopher Leibniz he was 'one of the most universal men that I know; nothing escapes his curiosity'.[38]

Thévenot's reputation went back to his activities in the service of the French crown – in the 1640s he had been part of the court of Louis XIII, before becoming French Ambassador to Genoa in 1647, and then to Rome in 1652. While in Italy he was involved in various shady dealings, including representing France in the long diplomatic negotiations that led to the election of Pope Alexander VII in 1655. In the 1650s he was also mixed up in the attempt by the Duc de Guise to launch a rebellion in Naples against Spanish rule. Then, in one of the most intriguing episodes of his diplomatic career, Thévenot 'received an order and a commission which the instructions I was given described as very dangerous and which I carried out with success and with the approval of Cardinal Mazarin' (Mazarin had become the real power in the country following the death of Louis XIII). In other words, he was on a highly sensitive and clandestine mission, as a kind of spy.

After Mazarin's death in 1661, Thévenot, by then in his early forties, turned his attention to what was to be his most influential work: the accumulation of books and knowledge. He had a huge library that became one of the founding collections of what is now the Bibliothèque Nationale de France. His popular teach-yourself-to-swim book, *L'Art de nager* ('The art of swimming'), was published posthumously and went into several editions in the eighteenth century, and was most recently reprinted in 1972. (In case the reader was unable to follow Thévenot's instructions, it also included handy advice on how to caulk a boat.)

A skilled linguist who spoke at least six languages, including English, Thévenot was particularly interested in the Arab world and the Orient, and he collected, translated and published manuscripts by travellers in a four-part series, each composed of several volumes, which are now highly valued by book collectors.[39] Thévenot frequently invited thinkers to his house on what is now the rue de Saintonge in Paris, not far from Montmor's home, or to his country house at Issy on the banks of the Seine, south-west of the capital, where experiments in physics, chemistry, astronomy and anatomy were carried out.[40] With the collapse of the Montmor *académie*, Thévenot regularised these meetings, holding them every Tuesday.[41] He installed a number of telescopes at Issy, from which Christiaan Huygens and others observed the rings of Saturn, and he intended to build a large observatory there. Huygens, who had been highly critical of the tendency of the Montmor *académie* to spend time on fashionable chatter rather than experimentation, was particularly grateful for the opportunity to do some real science.[42]

From 1664 to 1666 Thévenot financed these meetings and paid for the upkeep of visiting thinkers, all out of his own (deep) pockets. In the end, however, the cost proved too great and he wound up his informal academy, retiring to his house in Issy.[43] The day of the private scientific society was ending in France: the stage was set for the appearance of the Académie Royale des Sciences, with its complex administrative apparatus and its grandiose state-backed projects for 'big science', such as the construction of the huge Observatory on the edge of Paris.

Thévenot's vision of science was a fusion of experimental studies using scientific instruments (especially microscopes, telescopes and spirit levels), and geographical exploration, which involved both precision instruments and the discovery of new natural phenomena.[44] Some of his ideas seem pretty loopy to modern eyes – for example, he suggested that there was a continuous pulsation in the air which produced the movement of the heart, diaphragm and lungs, and he later proposed that the supposedly constant width of the cells in bee-hives should be used as a universal measure of distance.[45] However, an important part of early modern science involved sifting through apparently interesting natural phenomena by applying the newly established criteria of scientific investigation, in particular experimentation. Such ideas now look daft only because they were once studied seriously and were proved to be worthless.

The decision of Swammerdam, Steno and De Graaf to visit France was part of a long tradition of academic connections between the Netherlands and France. From the sixteenth century onwards, many Dutch students completed their studies with an 'academic pilgrimage' to France.[46] This was especially common amongst medical students, who would often gain their doctorate of medicine from one of the French medical schools. Caen University was particularly attractive because its doctorates were cheap and easy, to the extent that both sides of the transaction viewed the whole process with cynicism. The Caen professors had a saying: 'Take the money and send the dunce back home', while the Dutch joked that even a horse could get a degree from Caen, as long as it had the money. However, although Steno, Swammerdam and De Graaf all went to France at the end of their studies, only De Graaf had the classic motivation of getting a French doctorate – and not from Caen. In this, as in so many other things, Swammerdam and Steno stood apart from their younger friend and colleague.

This was not the first time Steno had come to Paris. Three years earlier, Thévenot had visited Leiden to pursue his oriental interests – the university library had a unique collection of oriental manuscripts, while the thriving Leiden printing industry specialised in Arabic, Hebrew and oriental fonts. During this visit, Thévenot invited Steno to come to Paris the following winter (1662–3), where the young Dane showed off his anatomical skills.[47] It seems probable that Thévenot also met Swammerdam on this visit to Leiden; De Graaf, however, was still in Utrecht at the time, and there is no evidence that he ever met Thévenot. Swammerdam and Steno's journeys to France, it appears, were made with the intention of attending Thévenot's *académie* rather than obtaining a doctoral degree – Swammerdam could not qualify as he had not yet finished his dissertation, while Steno had already obtained his doctorate. Swammerdam arrived in France first, travelling to Saumur, a beautiful town in the Loire valley, early in 1664. Although there were links between the Leiden medical faculty and members of the Saumur Académie (the Protestant university) which might explain why he visited the town, there is no trace of his name among the key students who enrolled at the Académie – indeed, there was no medical school there.[48] Instead, Swammerdam dreamily collected dragonfly larvae in the rushing waters of the Loire. To maintain his human-anatomy skills, he showed off to local physicians by demonstrating the dissecting techniques he had learnt at Leiden, revealing the delicate valves in the slender lymph vessels of the human body – something he had developed with Van Horne.

After a few months in Saumur, Swammerdam left for Paris, where he arrived in the summer of 1664. He soon met up with Ole Borch who was staying near the vineyards in a house on the rue Vaugirard on the south-western edge of the city, and over the next few months the two men saw each other regularly. As well as examining the latest additions to Swammerdam's collection of insects, they discussed a burn remedy which Swammerdam had used to treat himself when he had been injured by exploding gunpowder, various experiments by the Amsterdam chemist Catherina Questiers (one of the few female scientists of the time), and Descartes' techniques for studying magnetism. Above all, they carried out a number of extremely unpleasant experiments on dogs, which caused much pain, wrote Borch,[49] pursued a series of studies into penile erection which Swammerdam had begun in 1662, and investigated movement in frogs.[50]

Swammerdam also attended the lectures by the Cartesian philosopher and physicist Jacques Rohault, which were given every Wednesday at Rohault's house on the rue Quincampoix in the centre of the city. At one of these meetings, Swammerdam saw flies and frogs placed in an air pump and watched them gradually expire as a vacuum was created.[51] He also participated in another informal medically oriented discussion group, led by the Abbé Bourdelot, one-time physician to Queen Christina of Sweden, where he presented some recent findings in chemistry, including a recipe for sweetening sour wine (you add cinnamon oil and sulphuric acid – don't try this at home).

Most significantly, however, from November 1664 onwards, at around the same time as Steno arrived in Paris for his second visit, Swammerdam became a regular member of the Thévenot *académie*. Many of his contributions were similar to his discussions with Borch or the experiments he observed in the Rohault and Bourdelot *académies* – dissections and demonstrations of various chemical and medicinal procedures or substances, typical of the time and not in any way noteworthy. Steno's impact was far greater. As on his previous visit, he was an immediate hit with the Thévenot *académie* and beyond, carrying out dissections in front of a wide range of people, including professors from the University of Paris.[52] In April 1665 he left his mark on the history of anatomy and neuroscience when he gave his 'Lecture on the Anatomy of the Brain'.[53] Steno dissected the human brain in front of an audience of distinguished scientists and physicians at Thévenot's house in Issy, and provided the first thorough investigation of cerebral structure and of the organisation of the central nervous system. In itself, this was a remarkable feat, executed with great skill and, in the printed version, illustrated by some precise and highly informative drawings.

Steno's findings effectively demolished the main ideas of one of the key modern thinkers, Descartes. Although Descartes had died in 1650, his book *De Homine* ('On man'), which sought to explain brain function and human behaviour, appeared only in 1662, with a French translation appearing two years later, just as Steno arrived in France. In his presentation, Steno respectfully but repeatedly showed that Descartes' assumptions were completely mistaken. In particular, the ventricles of the brain, which Descartes argued were full of the 'animal spirits' he thought were involved in nervous transmission and muscular action, contained no such substance. Similarly, the pineal gland, which Descartes claimed was unique to humans and was the centre of the soul, was also present in other animals. By showing the superiority of experimentation over theorising, Steno struck the final blow in the long-running conflict between experimentalists and

Cartesian theoreticians which had destroyed the Montmor *académie*.

This period not only established Steno's international reputation; through Thévenot's initiative, it also altered the course of science itself. As Thévenot put it in a letter to Christiaan Huygens written in April 1665: 'We took the opportunity provided by the cold of recent months and applied ourselves to dissections and to investigating the Generation of animals.'[54] At first glance, the results of these initial

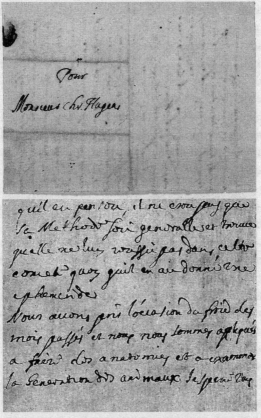

Thévenot's letter to Huygens, written in April 1665, announcing that his académie *had begun work on generation. At the top is the outside of the letter, showing how small it was folded, with the remains of Thévenot's seal at top left.*

investigations seem pretty minor. Thévenot invited Swammerdam to join his friend Steno at Issy, encouraging the two young men to work on a number of problems relating to generation. First, Swammerdam aided Steno in his research into generation in the chicken egg – 'that marvel of nature', as Swammerdam later put it.[55] Steno and Swammerdam's brief study[56] represented a continuation of research Steno had carried out the previous year on the role of the yolk,[57] together with a short description of the embryo's morphological changes. Although their results were relatively banal, mainly because they did not address the two most magical parts of generation – how fertilisation occurs, and how an organism develops from an apparently featureless egg – these studies marked the creation of a network of thinkers who studied generation.

Thévenot's decision to encourage his young protégés to study generation came shortly after the publication in French of Descartes' *De la formation de l'animal* ('On the formation of the animal'), which appeared for the first time in 1664, as a supplement to the French edition of *De Homine*. This unfinished fragment, written before Harvey's *De Generatione* was published, represented Descartes' attempt to resolve the problem of generation using his mechanical and mathematical theories. The starting point of Descartes' view was classic – a 'two-semen' Galen-style theory for mammals and birds, and spontaneous generation for insects. For Descartes, generation involved the action of Aristotelian 'heat' on the activity of 'particles' with different qualities. Set in motion by the 'heat' of fermentation, the embryo was created by particles that come either from a mixture of two semens or from decaying matter, argued Descartes. He set this view in an extremely ambitious framework: 'If we knew all the parts of the seed of some particular animal, man for example,' he claimed, 'we could deduce from that alone, on the basis of entirely mathematical and certain reasons, the entire shape and conformation of each of its members.'[58] However, there was a massive gap between his aims and his achievements – *De la formation de l'animal* merely contained some straightforward observations of embryonic development in the chick, described in terms that were neither striking nor novel. Overall, Descartes' attempt to describe generation in mathematical terms was a complete failure.

Nevertheless, in a Parisian intellectual environment still split between experimentalists like Thévenot and theoreticians inspired by Descartes, the publication of *De la formation de l'animal* was extremely important: it was a challenge to the experimentalists to do better. Through Steno's dissections, Thévenot's group was already proving that it could provide an explanation of brain structure and function that was far more accurate than that of Descartes. By turning the attention of his informal *académie* to the problem of generation, Thévenot was looking for more weapons in the continuing conflict over the shape and orientation of the future state-backed scientific society. If Thévenot did little at the time to publicise his *académie*'s work on generation, this was because the outcome was nowhere near as striking as Steno's work on the brain, and because Descartes' views were so vague and imprecise that they were not taken up even by his supporters within the Parisian scientific community. As Thévenot hoped, Descartes' theory-driven speculations were soon swept away by the rigorous empiricism of the new science, in particular by the work of the two young Leiden University students he had invited to his country house.

With this gentle push, Thévenot set in motion the great wave of studies of generation that changed the whole course of human understanding of our origins and development. Without Thévenot's encouragement it is unlikely that either Steno or Swammerdam would have made their later discoveries, and the generation network they helped to create would probably not have existed. Although it is possible that even without this vital step most of modern biology might not look very different – after all, someone, somewhere, sometime was going to make the kind of discoveries these young men made – the *way* we got to where we are today would have been subtly changed: the pace of discovery would have been altered, different mistakes would have been made and different false hypotheses would have been adopted. It is impossible to tell quite how different our understanding would be, but the four-dimensional shape of our knowledge would have been modified and we can be sure that the end product would be altered in all sorts of ways.

The lesson is that the history of science, like all history, is not written in advance. It is the result of the meeting of minds, of problems, of the material possibility of solving those problems, and of the specific, occasionally random conditions in which those problems are addressed. The question of generation, which had preoccupied humanity for so many thousands of years, had come to term. The techniques and concepts required for its resolution were now available, created by the growth of science and the economic and cultural changes that had spawned it. All that was required was inspiration and encouragement. That is what Thévenot provided in that cold winter of 1664–5; the turning point of this story, and of our history.

In summer 1665, as the heat intensified the smells of the city, Swammerdam and Steno left Paris. Swammerdam returned to the Netherlands, where he completed his doctoral dissertation and continued his work on generation in insects. Steno travelled slowly, first visiting Saumur and then spending several months in Montpellier, where he met up with a number of English students, including the naturalist John Ray, who had visited Leiden two years earlier, and the physician William Croune. At the beginning of 1666, while Steno was still in Montpellier, Chapelain tried to convince him to return to Paris – there were moves afoot to invite him to be a founder member of the Académie des Sciences – but he declined, and travelled on to Italy, taking the generation bug with him.

As Steno and Swammerdam moved on, Reinier de Graaf arrived in Paris.[59] He had left Leiden shortly after submitting his dissertation, and had then followed the traditional Dutch 'academic pilgrimage' to France. After qualifying as a physician at Angers in western France, which involved dissecting in front of the professors of Angers University – he studied the body of a sailor who had just been killed by a falling mast, tasting the still-warm pancreatic juice – De Graaf made his way to Paris. Around this time Thévenot withdrew to Issy, abandoning the Paris meetings of his *académie*. As a result, De Graaf never participated in the Thévenot *académie*, nor did Thévenot become his patron. Instead, De Graaf was befriended by Montmor, who was still interested in the latest medical discoveries, despite the recent collapse of his own *académie*. Montmor gave De Graaf a place in Parisian society, inviting him to dinner, taking him on trips into the country in his carriage and ostentatiously sitting him at his right hand during scientific discussions.[60] De Graaf's other patron was Thévenot's friend Chapelain, who helped the young man translate his books into French, and also introduced him to the medically focused group around the Abbé Bourdelot, which met in Paris every Monday.

Even had the Thévenot *académie* still been meeting regularly, it is not certain that De Graaf would have fitted in with its wide-ranging fundamental science agenda; his medical and therapeutic interests were far more in keeping with Abbé Bourdelot's applied outlook. Bourdelot considered that 'current medicine is worthless' and used his *académie* to encourage 'new remedies and new rules' in order to provide better treatment.[61] De Graaf's chemical study of the pancreas, and his recent invention of a flexible tube that could be attached to a syringe, making it possible for patients to self-administer an enema – much in vogue in both France and the Netherlands and soon to be satirised by Molière in *Le Malade imaginaire* – were signs of the modern approach to medicine that Bourdelot wanted to encourage.

During his stay, De Graaf dissected in front of the Paris medical professors, who were impressed with his dexterity but hostile to his modern ideas about anatomy, physiology and treatment. He also befriended another of Bourdelot's protégés, Jean-Baptiste Denis, a young French physician whose interest in blood transfusion soon landed him in court, charged with murder after one of his experiments unsurprisingly went wrong.[62] But as with Steno and Swammerdam, in retrospect the most important part of De Graaf's stay in Paris was his broaching the question of generation. In keeping with his medically oriented thesis and the interests of the Bourdelot *académie*, De Graaf's approach was clearly focused on human anatomy and medicine.

Encouraged by the interest shown in his dissections of the pancreas, he turned to the study of the male genitalia. Before leaving Paris in 1666, he dissected bulls' testicles in front of the Bourdelot *académie*. Although this minor work left no written trace, it was the first step on the road that led to De Graaf's major study of the male reproductive organs, which was published in 1668.

In different ways, the three friends had each been inspired by their stay in Paris. Steno had begun to think seriously about eggs, Swammerdam had refined his interest in insects to focus on the question of generation, while De Graaf had started his work on the anatomy and physiology of human generation. In the next six years their work would transform our understanding of reproduction and development, and would turn their friendship into rivalry and enmity.

3

INSECTS IN ITALY

Throughout history one of the great mysteries of the natural world has been the origin of small animals. Most people, most of the time, have believed the apparently obvious: they come from nowhere. This view held not only for tiny insects such as mites or maggots, but also for larger creatures, where belief in the lack of any lawfulness or coherence to generation could sometimes verge on the surreal – toads could be generated from ducks putrefying on a dung heap, a woman's hair laid in a damp but sunny place would turn into snakes, while rotting tuna would produce worms that changed first into flies, then into grasshoppers and finally into quail.[1] The people who believed these kinds of things were not necessarily fools. In discussing how scorpions could be generated by the smell of basil leaves that had been put under a brick, the alchemist Van Helmont dealt with the obvious suggestion that the scorpion had simply crept under the brick by claiming that the basil still generated scorpions if the leaf was placed between two heavy stones.[2] The resilience of the idea of spontaneous generation came from the fact that resolving many of these conundrums required more than casual observation and common sense.

Such stories were further reinforced by the fact that they were also found in ancient myths and literature. The Roman poets Virgil and Ovid explained that, to generate bees, all you had to do was to bury a bull with its horns protruding from the soil, wait for nine or thirty-two days (opinions differed), cut off the horns and bees would fly out. This story was still widely believed in early modern times – Pierre Gassendi, the seventeenth-century French physicist and atomist, accepted it without question, while in 1663 the Royal Society thought it would be a good idea for one of its members to follow this procedure.[3] There were at least three reasons why such stories were believed. Firstly, they generally came from the ancients and therefore should not be lightly abandoned;[4] second, they often appeared to have biblical support (in the case of bees, a variant on the bull story could be found in the Book of Judges (chapter 14, verse 8), where Samson saw 'the carcase of the lion: and, behold, there was a swarm of bees and honey in the carcase of the lion'); and finally – what other explanation was there?

Despite the weight of tradition, which was continually reinforced by everyday experience and common sense, by the early 1660s a number of thinkers had begun to wonder whether there was not another explanation for these stories, one more in keeping with the growing conviction that nature was governed by laws, which would exclude what Aristotle had called generation by 'accident'. In the late 1650s Robert Boyle, soon to be the leading light in the Royal Society, had outlined one important trend of thinking about the subject in an unpublished essay on 'spontaneous generation'. Boyle suggested that 'the wise Author of Nature' placed 'seminall principles' in each animal and that when the animal died, if the appropriate conditions were present, those 'principles' could 'be chang'd . . . into a Body of the texture requisite to exhibit such a determinate kinde of maggot or worme'.[5] For Boyle, generation occurred because heat activated these 'principles', which could then form life. This was similar to the position outlined by Descartes in his unpublished writings on generation, but Boyle cannot have known this. The parallels between their ideas flowed from the fact that both men had the same dual inspiration: on the one hand, Aristotle, and on the other, the growing influence of mechanical models of life. Boyle's view of generation was more innovative than that of Descartes: although, like everyone else, he argued that insects were generated from decay, he thought this did not occur by 'chance', but according to a set of rules – certain kinds of insect tended to emerge from certain kinds of matter. This was an attempt to reconcile common-sense observations with the spirit of the scientific revolution, without actually doing any experiments.

Given Boyle's interest in alchemy,[6] it is not surprising that his view of generation was also reminiscent of Paracelsus's recipe for generating a 'homunculus' by heating semen, and of the alchemist's wider vision of nature being full of potential life, according to which all things 'descend from the "Mysterium Magnum" which is the one mother of all things and of all elements and a grandmother of all stars, trees and creatures of the flesh'.[7] Although most of Paracelsus's views – in particular his suggestion that frogspawn was 'the sperm of the world', responsible for generating insects, making plants green and so on – were by this time largely discredited,[8] the parallels with Boyle's ideas are striking.

The Royal Society itself paid relatively little attention to the question of generation in its early years, being far more concerned with making an air pump, measuring the Earth, weighing the air or even discussing Sir John Finch's invention of an incombustible hat band.[9] Unlike at the Thévenot *académie*, generation was not something that the London thinkers considered to be of decisive importance. But even though their discussions were very limited, they still revealed the depth of contemporary confusion on the subject.

Just like everyone else, the members of the Royal Society could not explain where organisms came from, and were not even sure that like bred like. They were happy to listen to tall tales such as that of 'a maid in Holland who voided seed by urine, which being sown grew',[10] or of 'a copulation of a male rabbit and a female cat, which produced monsters, whose foreparts were like a cat, and the hinder parts like a rabbit; and that those monsters had reproduced more complicated monsters'.[11] While it is possible that these stories were greeted with amused laughter, there is no doubt that in May 1661 the Royal Society listened with the greatest care to the suggestion that it was possible to arrange 'the production of young vipers from the powder of the liver and lungs of vipers'.[12] They took the claim so seriously that they decided to investigate it, returning to the question repeatedly over the next two years. Having eventually procured 'a glass-jar, full of the powder of the bodies of vipers, and a gallipot full of the powder of only the hearts and livers of vipers', the contents of which were checked on a number of occasions, they finally had to accept in June 1663 that the viper powder had turned into a smelly mess, and that while it was full of 'little moving creatures', there were no vipers.

Even the more rational comments made at the Royal Society's discussions of generation reveal what remained to be discovered – on 24 September 1662, George Ent, who by this stage was a leading Fellow, declared that 'it was found by experience that no oaks grow well but from acorns'. Assuming this was not a joke, the very fact that it needed saying shows that not everyone was convinced it was true. At a discussion that took place six weeks later, someone (probably Boyle) took the idea of 'seminal principles' a step further, suggesting that they might be airborne – 'where the animal itself does not immediately furnish the seed, there may be such seeds, or something analogous to them, dispersed through the air, and conveyed to such matter as is fit and disposed to ferment with it, for the production of this or that animal'.

The Royal Society recognised that experimental proof was needed to resolve the question, so they commissioned a series of studies on the generation of insects. Unfortunately, none of the 'experiments' came to very much, partly because they were not clearly thought out, partly because, despite three Fellows agreeing to put 'blood, flesh, brains etc together in a glass or other proper vessel; as also bran and meal; and likewise cheese moistened with sack, etc.', the trio apparently did nothing for six months, and another three men were eventually given the task in March 1663. This time John Evelyn 'undertook particularly to put several pieces of flesh and some blood in a closed vessel, which might not be fly-blown, to see what it would produce'. Not surprisingly, he came back a couple of months later to report that the concoction had 'bred nothing'. The matter was finally closed in July 1664 when 'Mr EVELYN gave an account, how that flesh, which he had formerly, by order of the Society, put in a glass, covered with double flannel, had bred no live creatures, but turned into a mucilage, and then dried up'.

Nothing came of this report, probably because it was not obvious exactly what the experiment meant and what conclusions could therefore be drawn about generation. While the implication was that meat that was not 'fly-blown' would not generate insects, it was not clear why – perhaps the 'double flannel' had prevented Boyle's airborne 'seminal principles' from coming into contact with the meat, or perhaps the mixture had not been sufficiently warm for these 'principles' to be activated, and so on. Boyle was clearly aware of this problem, for the following year he announced to the Royal Society that he had set out a 'catalogue of experiments relating to spontaneous generation', although he gave only two examples of how he would present meat to see if it would generate insects – '1. in glasses hermetically sealed, having the ordinary air in them; 2. in glasses first exhausted and then sealed up'.[13] This would at least have shown whether air was necessary for generation, but Boyle's suggestions were never taken up.

This somewhat lackadaisical attitude was typical of the Royal Society's occasional interest in generation during the period 1660–65: their unfocused musings had neither impact nor consequence. This lack of rigour was specific to generation – the Society's work on air pressure, for example, was much more serious and was consistently followed up. But there was also a general problem: in the early years of the Royal Society's existence, public diffusion of its work was hampered because it had no public face – its journal, the *Philosophical Transactions*, was first published in 1665.[14] As a result, in the early 1660s no one outside the small circle of men present at the meeting knew what had been discussed.[15]

The Royal Society never seriously returned to the issue of generation; its work was hampered in 1666 first by an outbreak of the Plague, then by the Great Fire of London which destroyed its headquarters and disrupted its work for many months. During this time, the Royal Society's amateurish attempts to deal with the subject were swept away by the clash of two completely contradictory approaches which were published in Italy, one largely based on hearsay and fantasy, the other a model of scientific rigour, directly inspired by Thévenot's circle.

Tuscany is rolling, fertile country, stretching from Livorno on Italy's west coast up into the Apennines, with the Chianti vineyards at its heart. In the late medieval period most of the Tuscan cities became independent communes or republics, as princes were ousted by merchants and bankers who demanded political power to match their economic influence. Growing wealth, together with an intellectual outlook that fed and accompanied a wave of political upheavals, transformed these city states into key centres of the cultural revolution that became known as the Renaissance.[16] Two centuries later, Tuscany would play a similar role in the scientific revolution.

By the middle of the sixteenth century the Medici family, who had come to rule Florence when the republican experiment failed, had extended its power over the whole of Tuscany. Through their financial influence with the Papacy, the Medici had succeeded in transforming themselves from a mere banking family into a *nouveau riche* clan of ambitious aristocrats and calculating cardinals. The Medici fusion of political, religious, cultural and economic power was not unique, but they were undoubtedly the most successful, powerful and influential family of the period. In the fourteenth and fifteenth centuries, they encouraged artists like Botticelli and Michelangelo and unclassifiable geniuses such as Machiavelli and Leonardo da Vinci.[17] The fact that the Medici – and Florence – were associated with such an astonishing set of people gives some indication of the power and influence of the city and its ruling elite.

By the seventeenth century, the Medici's reputation had begun to fade in proportion to the family's declining wealth. They started to wind down their banking activities, and their fortune increasingly flowed directly from the coffers of what had become the Grand Duchy of Tuscany. This may have been a miscalculation: state revenues fell as the first half of the century was marked by a general decline in European trade, partly prompted by population crises (there was an outbreak of the Plague in the 1630s), and partly due to the wars that wracked the Continent. Although the Italian states were free of major armed conflict during this period, they went through an economic crisis in the early decades of the century. Tuscany was particularly hard hit, with a 60 per cent decline in wool production between 1620 and 1660. Florence's loss was Leiden's gain, however – the guild regulations which restricted economic activity in Tuscany were weaker in the modern economy of the Dutch Republic, and new techniques and new products were readily adopted, enabling the Netherlands to outcompete the old producers to the south.[18] As a result of all these factors, the Medici fortune continued to decline. Over the course of the century the power and influence wielded by Tuscany, and by the Medici in particular, gradually dwindled.

The failing power of the Medici was the backdrop to the most tragic event in the whole of the scientific revolution: the crushing of Galileo by the Catholic Church. At the beginning of the seventeenth century, Galileo was a brilliant but penniless mathematics lecturer in Pisa who seized upon the invention of the telescope in the Netherlands to discover fundamental truths about the Universe, and to secure himself a better job. In 1610 he published *Sidereus Nuncius* ('Starry messenger'), an astonishing book in which he showed pictures of mountains on the moon as seen through his telescope, and described a vast number of new stars, including four moons orbiting Jupiter. In a moment of inspiration, he called the new moons 'the Medicean Stars' after the new head of the family, Grand Duke Cosimo II, and his three brothers. Galileo clearly hoped to gain the support of the Medici – preferably financially (this was not a complete shot in the dark, Galileo had briefly taught Cosimo in 1605 when the young prince was an idle teenager).

Galileo's ingratiating gesture paid off, and Cosimo II appointed him 'First Philosopher and Mathematician' to the court in Florence. For Galileo this was a huge step forward; being a courtier brought him security and status, and for his salary he was simply expected to pursue his research – no teaching, just study, and plenty of telescopes. The capital of Tuscany, with a population of around 75,000, was far smaller than Paris, London or Amsterdam, and the days of Dante and Michelangelo were long gone, but the force of Galileo's ideas and discoveries put Florence back at the centre of Europe's intellectual map. At the same time, however, Galileo found himself in mortal peril.

Cosimo II died in 1620 and was replaced by his ten-year-old son, Ferdinando II. It was during the early years of Ferdinando's regency that Galileo refined his understanding of the solar system, and embraced the view of the Polish astronomer Copernicus, who had argued that the Earth goes round the Sun, against the common-sense view that the Sun goes round the Earth. This theory had been condemned by the Church, but Galileo was convinced it was true. In the meantime, Ferdinando grew into a plump, charming bisexual who enjoyed hunting, fishing and bowls, but who was also interested in science and intellectual debate.

Five years after coming of age and taking control of the throne, Ferdinando had a decisive opportunity to show his independence and his attachment to the importance of scientific discovery. In 1633 the Pope summoned Galileo to account for the positions taken in his latest book, *Dialogo sopra i due massimi sistemi del mondo* ('Dialogue on the two chief world systems'), which had been published in Florence with Ferdinando's approval in 1632. By this stage, Galileo was a frail sixty-eight-year-old and his eyesight was beginning to fail. If the old man expected some leniency from the Pope, who had been his patron in the past and had supported him in previous, less decisive disputes, he was to be sorely disappointed. The Church now persecuted him as an example to all those who questioned its authority, threatening him with torture and death, forcing him to deny what he knew to be true, and condemning him to permanent house arrest. Ferdinando, his patron and employer, was unwilling or unable to prevent any of this: if he were to voice his support for his scientific servant he would be risking war and excommunication in defence of Galileo's ideas – not a very likely proposition.

As Galileo must have realised, the possibility of Ferdinando coming to his aid was effectively reduced to zero by the fact that relations between Tuscany and the Papacy were already extremely tense. The year before, papal troops had occupied the Duchy of Urbino, despite Ferdinando's legitimate claim to the territory. Faced with the prospect of war, Ferdinando had been prudent and had allowed the Pope to get his way. At the opening of proceedings against Galileo, the Pope had warned Tuscany not to interfere. Again, Ferdinando did nothing, although he did allow Galileo to stay with the Tuscan Ambassador to Rome during his trial, and to return to Tuscany for his house arrest, which lasted until his death in 1642.

Through its persecution of Galileo the Church showed quite how seriously it took the threat of science, in particular anything that might contradict Holy Scripture, or undermine the fusion of Aristotelian philosophy and Christian theology which had held sway in Europe for more than three hundred years. Nevertheless, Italian science was neither stifled nor condemned to study marginal anodyne topics, and in the following decades Florence was the centre of some key discoveries which were of decisive importance for our understanding of generation.

Ferdinando may not have covered himself in glory during the persecution of Galileo but he did maintain his interest in science. In the years following Galileo's incarceration, Ferdinando invented a thermometer using glass balls full of spirit, as well as several devices for measuring humidity. And in 1656 he invited one of the rising stars of Italian scientific anatomy, Marcello Malpighi, to be Professor of Theoretical Medicine at Pisa medical school. Beneath Ferdinando's jovial, courtly manners there was a keen awareness of the conflict between Aristotle's teachings and the discoveries that thinkers were making throughout Western Europe. He was particularly concerned about the influence of the followers of Aristotle in schools and universities. This was the background to a huge row between Ferdinando and his pious and portly wife, Vittoria, when the couple disagreed over the education of their son, the future Cosimo III. Sadly for Ferdinando (and Cosimo), Vittoria won the argument and entrusted her son to the Jesuits, who had little difficulty in transforming the boy into the opposite of his father, a sad child with an unhealthy interest in martyred saints.

While Ferdinando might have lost the battle of modernity versus tradition on the domestic front, he had greater success when it came to the court. In 1657, together with his brother Prince Leopold, a free-thinking priest and future cardinal, Ferdinando set up a small, informal and largely private scientific society, the Accademia del Cimento (Academy of Experiment).[19] The Accademia met irregularly in private, and had neither a constitution nor a publication. Hidden from the public gaze and devoted to a mixture of court entertainment and genuine scientific discovery, the Accademia attempted to find answers to a range of key physical questions – mainly to do with the existence of a vacuum and the nature of cold.

With its motto taken from Dante, 'Provando et reprovando' ('Test and test again'), the Accademia had less than a dozen participants, including Ferdinando and Leopold, who would watch experiments being carried out. Having two of the richest men in Italy as patrons meant that the Accademia wanted for nothing, with a well-equipped laboratory in the Pitti Palace on the south bank of the Arno. It was much more than a source of aristocratic amusement, however, and from the outset its proceedings were the scene of serious conflicts between Aristotelians and modern thinkers over the meaning of the experiments they were carrying out.[20] These battles would frame both the work of the Accademia and the presentation of its findings to the public.

The Accademia's only official was a young courtier, Lorenzo Magalotti, a poet, mathematician and dancer who acted as the

Secretary and had the unenviable task of writing up accounts of the various experiments.[21] This took many years and was doubly difficult: each of the participants wanted to check what Magalotti had written, and he had to steer a careful course between the moderns and the Aristotelians, avoiding any possible offence to the Church while faithfully reporting the results of the experiments. This may help to explain the absence of any speculation or theory in the accounts of the work of the Accademia, as well as the use of the third-person passive voice ('it was seen that'), a linguistic device which is still used in today's scientific articles to reinforce their objectivity.

The results of the Accademia's experiments, virtually all of which dealt with physical phenomena such as air pressure and temperature, were finally published in 1667, in a 260-page book entitled *Saggi di naturali esperienze* ('Examples of experiments in natural philosophy'). Like Galileo's key work, the *Saggi* were written in Italian, not Latin, suggesting that they were intended to reach a wider audience than Latin-speaking scholars.[22] However, there was no question of selling the book – the Grand Duke could not be associated with something as vulgar as commerce, no matter how his ancestors had made their fortune. Instead, a few hundred copies were distributed amongst the great and the good[23] and that, more or less, was that. Shortly afterwards the Accademia was wound up for reasons which remain unclear – perhaps it had simply completed its work. To set the seal on its demise, Leopold became a cardinal and moved to Rome, while the *Saggi* lay in relative obscurity until the book was taken up by a commercial publisher twenty-five years later.

The Accademia was very different from the Royal Society, and from the various French *académies* of the time. Despite its name and its royal charter, the Royal Society had little to do with the throne – neither Charles II nor any subsequent ruler paid it any serious attention or money. The Society had a clear structure, a public face and role, and from 1665 onwards it had an outlet for its work, in the shape of the *Philosophical Transactions*. By contrast the Accademia was unofficial and private – far more private than the Thévenot *académie* or any of the other French groups. Although vague reports of its extremely irregular activity circulated in court and intellectual circles throughout Europe, its immediate influence was very limited. The Accademia took on this very private form precisely because, unlike all other scientific societies at the time, it was intimately linked with the everyday life of the court. Ferdinando and Leopold not only bankrolled the Accademia, they also participated in many of its activities. Because their status was threatened when the royal brothers appeared on something approaching an equal footing with ordinary men, the life of the Accademia became part of the court, and some of the scientific writings of those associated with it are now seen as being as much products of courtly games and rhetoric as part of the new way of understanding the world that would come to be called 'science'.[24]

This mixture of royal entertainment and a genuine attempt to fashion new, objective knowledge was particularly evident in the work of Francesco Redi, physician to Grand Duke Ferdinando II of Tuscany, Keeper of the Royal Pharmacy, courtier and part-time poet.[25] Most of Redi's investigations were performed outside of the framework of the Accademia – he carried on experimenting and publishing for nearly thirty years after its demise – but they were all done with the Grand Duke's support, either tacit or explicit. In the mid-1660s, in the glorious surroundings of the Tuscan hills, Redi carried out a series of investigations into the generation of insects which showed the full power of the experimental method, and which still amaze today because of their audacity and thoroughness. The importance of what he was doing lay not only in what he had found out, but perhaps even more in the *way* he had found it out. Like a true courtier, Redi employed wit and erudition as he presented his findings and elegantly demolished the arguments of his opponents, using their own fables and fairy tales to highlight their errors. But behind the refined humour and genteel intelligence there was a determination and rigour that marked him out from the rest of the court, and indeed from most other thinkers of the time. Despite his urbane airs, Redi was an astute experimentalist whose insights resonate down to the present day. Through his work on insects, he made a key contribution to our understanding of generation and what it means to do an experiment. Redi's inspiration came from two sources: his close friendship with Steno, who transmitted the Thévenot circle's interest in generation, and his profound irritation at the old-fashioned ideas published by another Italian thinker.

In 1633, shortly after the papal court sentenced Galileo to permanent house arrest, a German Jesuit priest named Athanasius Kircher (pronounced 'Keer-ker'[26]) arrived in Rome to take up the post of Professor of Mathematics and Oriental Languages at the Roman College. Over the next half-century Kircher not only supervised the College's fantastic collection of 'curiosities', which attracted visitors from all over Europe, he also published an amazing number of books. As 'the last man who knew everything',[27] he wrote about an incredible variety of subjects, from Ancient Egyptian hieroglyphics and antiquities, through the structure of the Earth and stars, the nature of music and the usefulness of codes, to a bewildering array of monsters and mythical beasts.

Although Kircher stands out as one of the publishing phenomena of the seventeenth century – he produced nearly forty books, including many large astounding volumes with beautiful engravings – his reputation amongst his contemporaries was very mixed. His readers were impatient to get hold of his books and clearly treasured them for their beauty. But as the century wore on, thinkers became increasingly wary of Kircher's intellectual standards, openly laughing at his amateurish mathematics, ridiculing his gullible appetite for stories about dragons, and even setting him up by presenting him with spoof 'ancient' manuscripts and then chortling at his comments. Their private views of him were damning: for Descartes he was 'more charlatan than scholar', one of Redi's Roman correspondents said he was 'highly susceptible to suggestion', while the German philosopher and mathematician Gottfried Leibniz stated bluntly: 'he understands nothing'. It was simply not possible to believe what

Athanasius Kircher.

Kircher wrote: having failed to replicate one of his reports about the effect of concentrated moonbeams on saltwater, Henry Oldenburg, the Secretary of the Royal Society, complained that 'ye very first Experiment singled out by us out of Kircher, failes, and yt 'tis likely, the next will doe so too'.[28]

Despite the fact that many of Kircher's ideas were hilariously wrong and he clearly was a very credulous man, his enormous popularity amongst seventeenth-century readers shows us how European society viewed knowledge and the way it should be obtained. For most people, there was nothing necessarily absurd about Kircher's concoctions of travellers' tales, myths and wild speculation. This was the way folk knowledge was transmitted, and that most revered source of knowledge, the Bible, was full of such stuff.

Furthermore, new scientific techniques and the exploration of the New World had led to the discovery of the most astonishing natural phenomena. There was no reason for most people to disbelieve something they read or heard just because it was out of the ordinary. However, that did not apply to everyone, and some of Kircher's contemporaries were irritated by his writings, mainly because they thought he was wrong but partly, perhaps, because they were jealous of his publishing success which brought not only fame but also, presumably, wealth.

In 1665, Kircher published one of his most stupendous books, *Mundus Subterraneus* ('The subterranean world'). Six hundred pages long, divided into twelve parts, each packed with sumptuous engravings, and retailing for the princely sum of fifty shillings,[29] *Mundus Subterraneus* was Kircher's collection of ideas about everything to do with the Earth. He was particularly interested in the nature of volcanoes (he had been lowered into the smoking crater of Vesuvius in 1638), which he thought were connected to a fiery furnace at the centre of the planet, providing evidence for his suggestion that the Earth was a star that had gone cold. He also presented his theory of tidal movement, studied the nature of fossils,[30] reproduced some delightful pictures of dragons, described races of humans that lived underground and, in the final part of his book, discussed the generation of insects.

Kircher's view of generation was not original, but it had the enormous virtue of summing up classical, medieval and early modern ideas about the question. Like most modern writers, he claimed to be basing his ideas on experience and experiment. But for Kircher 'experience' meant accepting eyewitness accounts from modern travellers or from ancient manuscripts, while an 'experiment' might involve merely a single attempt to produce an effect, without trying to use the results to analyse why the effect occurred. Kircher not only took unreliable people at their word, he then compounded matters by turning vague descriptions into precise instructions for generating a variety of organisms – snakes, flies, scorpions, bees, silkworms, and frogs.

If you wanted to generate flies, Kircher's procedure was sure to succeed: 'Collect a number of fly cadavers and crush them slightly. Put them on a brass plate and sprinkle the macerate with honey-water. Then expose the plate, as chemists do, to the low heat of ashes or of sand over coals, or even of horse dung; and you will see, under the magnifying power of the microscope, otherwise invisible worms, which then become winged, perceptible little flies, and increase in size to animated full-fledged specimens.'[31] For Kircher this and similar accounts showed that nature 'is impregnated by the power of the circling stars and mixtures of seminal principles'[32] – these 'seminal principles' were made up of the key alchemical compounds (salt, mercury and sulphur). Just add heat and water, and life would appear. Although Kircher made a link between this triad of magical substances and the Holy Trinity, the real source of his view, as with Robert Boyle, was the alchemical vision of Paracelsus and his 'Mysterium Magnum'.

This was the problem with Kircher. Leaving aside those parts of his material that were simply fantasy (dragons, men living underground), much of what was true (you can indeed generate flies using his recipe) was placed in a framework that made it useless for further study. The point is not that you can generate insects from rotting matter, but rather what that means – how the insects got there. For Kircher, it was proof of a concept that flowed from a semi-mystical, pre-scientific approach to the natural world. His observations might have been more or less accurate, but they were not sufficiently precise to enable him to challenge and test his hypothesis that nature is full of 'seminal principles'. Had he watched a bit more closely, he would have realised that rotting matter is in fact just full of flies laying eggs. Proving that point, however, was more difficult than might appear.

August 1666. A thin man stood in the large oval courtyard in front of the Grand Duke of Tuscany's palace at Poggio Imperiale, just south of Florence. He wore a smart coat, fashionable white stockings, and was staring at a wooden frame covered with thin cloth. He gestured lazily to a servant, who knelt down, rummaged around and brought out a large glass vase, its mouth covered with the same thin gauze, and placed it on the ground in front of his master. The thin man crouched down, feeling the sun beating on his back; sweat trickled on his scalp, making his wig itch even more. He lifted off the gauze, reeled back at the stench and clutched a piece of silk to his face. Taking a long thin piece of wood he poked around in the fetid mess inside the vase. After a few seconds he grinned triumphantly and gestured to his friend Steno, summoning him to come and look. There was nothing there!

The thin man was Francesco Redi and he was engaged in an unprecedented experiment to test some of Kircher's wilder claims and resolve the puzzle of the generation of insects. These investigations

Francesco Redi.

were important not only because of their subject matter but also because they were real experiments, designed to decide between alternative explanations. Another link in the chain of modern science was being forged under the bright Tuscan sky.

The son of a physician, Redi was born in the city of Arezzo in 1626. With his elegant style and urbane wit, he was a perfect addition to the court, observing at least some, and probably most, of the Accademia del Cimento's experiments and debates.

In 1664, Redi had published a study of the venomous power of vipers, *Osservazioni intorno alle vipere* ('Observations on the viper'),[33] a witty, courtly public letter some 12,000 words long, written in Italian to his friend Magalotti, the young Secretary of the Accademia. As Redi told it, his experiments on vipers were effectively a royal command performance, following a lively discussion between members of the court about the nature of viper venom: 'Matters were growing disputatious when H.R.H. [Ferdinando] commanded that every experiment be done to find the truth, that everyone should retest his opinion as he see fit.'[34]

In his first book, Redi described how, together with other members of the court, he tested various possible sources of the venomous power of the viper. Having examined more than three hundred, and studied the effect of venom on a substantial number of birds and other creatures (including a footman), Redi decided that the poison was a yellow substance produced in glands in the snake's mouth, and that it was injected into the skin through 'sheaths' around the reptile's fangs. Furthermore, he concluded that to act as a poison, venom had to come into contact with the blood – it could therefore be swallowed quite safely. This was dramatically demonstrated when, to the amazement of the assembled courtiers, Jacopo Sozzi the Viper-Catcher knocked back half a glass of wine spiked with a massive amount of viper venom 'as if it had been so much pearly julep'. It is striking that Redi did not use himself as a guinea pig, although whether this was out of prudence, or to maintain the dignity of his station, is not clear.

As this shows, Redi not only carried out precise and logical scientific studies, he also produced an entertainment – for the court and for his readers, who were not necessarily presumed to be scholars. His witty, wry, self-deprecating style flows across the centuries, giving the impression of a highly cultivated, likeable man. Redi claimed that he was trying to uncover the truth about viper venom using a new, different way of finding out about the world: 'I love Thales, I love Anaxagoras, Plato, Aristotle, Democritus, Epicurus, and all the princes of philosophy', he wrote, 'but I do not, however, wish servilely to bind myself to swear that all they have said or written is true.' His approach was apparently straightforward: 'Every day I find myself more firm in my intention of not trusting the phenomena of nature if I do not see them with my own eyes and if they are not confirmed by iterated and reiterated experience.' This was partly rhetoric. Redi – like everyone else at the time, before and since – had to take other people's word for it about some things. But unlike Kircher, Redi took the Accademia del Cimento's motto to heart, and accepted the need to test and retest his findings and, above all, his ideas.

In retrospect, Redi's work on the viper was just as striking for what it did *not* contain. While the Royal Society was busy trying to generate vipers from powder, Redi simply ignored the question of generation. Indeed, up until 1666 there is no evidence that any of the scientists who filled the Medici court were interested in the subject. But after that date, the surroundings of the Pitti Palace and the Villa Poggio were full of rotting flesh, buzzing flies and wriggling maggots. There were two reasons for this change in outlook. In 1665, Kircher's monumental *Mundus Subterraneus* appeared, full of fantastic drawings and lazy thinking. From the way Redi later dealt with Kircher, there is little doubt that he was provoked to test the Jesuit's claims. More decisively, however, in March 1666 Steno arrived from France, bringing with him news of the work on generation being done at the Thévenot *académie*. Redi and Steno struck up a very close friendship and for the next two years they became inseparable, socialising and working together. As Redi put it in a letter of March 1667: 'Signor Steno does me the honour of favouring my table day and night, and

I am content to enjoy his most learned and amiable conversation.
What is more, we are never at leisure, dissecting every day and making
observations.'[35] During this time the two friends inevitably discussed
the problem of generation and how to study it. But although the idea
of investigating the generation of insects, and even certain experiments, may have come from Steno (he definitely participated in some
of Redi's observations), the form and rigour of the work on insect
generation was far closer to Redi's previous research on vipers than it
was to anything Steno had ever done.

Redi published his findings in 1668, under the title *Esperienze
intorno alla generazione degl'insetti* ('Experiments on the generation of
insects').[36] Like *Osservazioni intorno alle vipere*, this book took the
form of a letter in Italian, this time addressed to his friend and fellow
courtier Carlo Dati.[37] Its style was typical Redi – lively and relaxed,
full of wit and sophisticated references to classical and contemporary

*Title pages of two editions of Redi's book on the generation of insects. The
original Italian edition (left) and the later Latin edition (right).*

poetry. And, unlike most such publications, it did not contain any fawning references to the author's patron (Ferdinando).[38] Although Redi was above all a courtier, and his book was designed to amuse and instruct the court, such was his confidence and status that he did not feel the need to ingratiate himself. This book was also completely different from the *Saggi*, with its dry and bitty descriptions of experiments on physics. Redi's work dealt with a problem of natural history, described in a conversational and cultivated style, coming to a radical conclusion through a series of logical steps, each backed by experimental evidence and classical references.

From the very first page, Redi nailed his flag to the mast by reproducing a telling Arab proverb: 'Experiment leads to knowledge, credulity leads to error.' Given that Kircher repeatedly pops up as a fall guy in the pages that follow, this can probably be taken as a sideswipe against the gullible Jesuit.[39] Redi began his book by summing up previous views of insect generation, starting with pagan Greek philosophers such as Epicurus and Democritus, who thought the 'Earth Mother' had at first regularly generated humans and large animals from the soil, but eventually lost this power, although she was still able to produce 'small creatures such as flies, wasps, spiders, ants, scorpions, and all the other terrestrial and aerial insects'.[40] Redi then outlined the modern Christianised Aristotelian version of this view, whereby the Earth possesses some 'generative principle' that remains alive and is awakened by heat, resulting in the generation of insects.

Redi contrasted this orthodoxy with his own radically different vision: 'the Earth, after having brought forth the first plants and animals at the beginning by order of the Supreme and Omnipotent Creator, has never since produced any kinds of plants or animals, either perfect or imperfect; and everything which we know in past or present times that she has produced, came solely from the true seeds of the plants and animals themselves, which thus, through means of their own, preserve their species'. In the case of insects, Redi boldly claimed that they were 'generated by insemination and that the putrified matter in which they are found has no other office than that of serving as a place, or suitable nest, where animals deposit their eggs at the breeding season'.

To prove his point, Redi did an experiment. On a June day in either 1666 or 1667 he had three snakes killed, and placed them outside in an open box. After a few days the dead animals were covered with 'worms', but Redi could not draw any conclusions as to what these worms were, because they escaped 'through a hole in the box'. Undaunted, he plugged the hole and did the experiment again. This time he found the worms eventually turned into red or black 'eggs' from which emerged either blowflies, 'green-bottles' or tiny black flies.[41] Surprised to find that snake meat had produced more than one kind of fly, Redi did the obvious thing – another experiment.

This time he took six open boxes, each containing bits of snake, pigeon, veal, horse flesh, chicken meat or a sheep's heart. The same thing happened, except this time he got blue and violet flies as well. The worms, he thought, came from the small white eggs he observed on the dead flesh. To see whether different types of fly were generated from different types of meat, Redi did the experiment again, and again, and again, using an astonishing array of dead animals – ox, deer, buffalo, lion, tiger, dog, lamb, kid, rabbit, ducks, geese, hens, swallows, swordfish, tuna, eel, sole, 'etc'. Each time he found the same thing: the eggs produced 'worms', which turned into what Redi called 'eggs' of certain colours (these were what we would call pupae), which then produced specific kinds of flies. But although all this work demonstrated that there was a clear link between each stage in the insects' life cycle, it did not show what that link was, nor where the original eggs came from.

The next step in Redi's experiment took as its starting point not a scientific insight, but an episode from classical poetry. In Book XIX of Homer's *Iliad*, Achilles fears that flies might breed worms in the wounds of his dead friend Patrocles. To investigate exactly what Homer meant, Redi took eight large flasks and put bits of snake, fish, eel and veal in them. Half of the flasks were left open to the July air, while the other half were covered with lids made of paper, tied tightly with string. As you might expect, Redi found maggots and flies in the open flasks, but nothing in the closed flasks, although he did notice that sometimes there would be a maggot on the outside, trying to get in. He then did the same experiment at different times of the year, using different containers and different meats, but always got the same result. He even found the same thing happened if he replaced the meat with crushed maggots or dead flies: if the container was open, he got flies; if it was shut, he got nothing.

That could have been the end of the story, but Redi wisely realised that there was a problem with his experiment: air could not circulate in the closed containers. If airborne 'seminal principles'

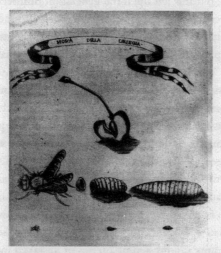

Redi's illustration of the stages in the life of the 'Cherry fly'.
The maggot is on the right, the pupa in the centre.
At the bottom, Redi has drawn them life-size.

generated fly eggs, they would not be able to enter his apparatus. Or again, perhaps the worms required fresh air in order to grow. So he did the experiment once more, taking a large container and covering the open top with 'a fine Naples veil' or piece of gauze. To be doubly sure that no flies could enter, he put the covered container in a large gauze-covered frame, a kind of meat safe. Once more, there were no maggots or eggs to be found on the meat. However, he did notice many flies outside the net, some of which laid live maggots. He then dissected a number of flies (no mean feat) and noted that flies have two ovaries that could contain up to two hundred eggs each. Putting all this information together, Redi correctly concluded that 'some kinds of flies bring forth live worms and some others eggs'.

This might seem to have involved enough rotting meat and buzzing flies to satisfy anyone's curiosity, but Redi was just getting into his stride. He had demonstrated the origin of maggots found on rotting flesh, but this left open the generation of insects from other forms of decay. So he showed that the worms found in cheese[42] came from fly eggs, which produced skipping maggots, which then turned into pupae, from which hatched male and female flies which would then mate. He also did similar experiments on vegetable matter, and succeeded in finding eggs (and then maggots, then pupae and then flies) on mashed melon, cucumbers, strawberries, pears, apples, plums, lemons, figs, peaches, raw and cooked pumpkin.

Throughout his book, Redi contrasted his findings with the arguments of previous writers, from Aristotle to Gassendi. But the man who had summarised the greatest variety of mistaken material on insect generation was Kircher, so Redi returned over and over again to his views. For example, he took Kircher's honey-water recipe for generating flies – 'a single, ill-considered experiment', scoffed Redi – and showed how it consisted of a series of mistaken interpretations: the honey-water would attract live flies which would then lay eggs, you got the same result without the brass plate and low heat, while Kircher's description of the maggots turning straight into flies was simply wrong. In a particularly sharp aside, Redi underlined the link between Kircher's views and unhealthy alchemical ambitions, pointing out that the Jesuit's account 'must have delighted and elated those persons who fondly imagined that they could re-create man from man's dead body by means of fermentation, or other similar or still more extraordinary processes!'[43]

After following Kircher's instructions for generating 'caterpillars' out of ox dung, which were then supposed to turn into bees, Redi reported he found only flies and gnats, and that if he took the elementary precaution of shielding the dung from egg-laying insects, he observed 'no generation of any kind'. When he tried generating scorpions from rotting scorpions, as Kircher suggested, 'I always got flies,' Redi sighed. And faced with Kircher's claim to be able to generate frogs from the 'fertile dust' to be found in a sun-baked swamp, Redi remarked drily, 'This experiment ought by all probabilities to succeed, but I have never had the honour of being able to confirm it, owing possibly to some lack of attention on my part, or to some unknown obstacle, which, however, may be found in my having carried out Father Kircher's rule to the letter.'

Kircher was not the only object of Redi's wit. Confronted with the great medieval Arab philosopher Avicenna's surreal assertion that 'a scorpion will fall dead if confronted with a crab to which a piece of sweet basil has been tied', Redi said with a weary shrug: 'it is likewise false, and having proved it so, I passed on to further experiments'. Even Ferdinando's court was occasionally the butt of Redi's humour – in the best possible taste, of course. While he and Steno were staying at the Villa del Poggio Imperiale, in summer 1666 or 1667, the two men noticed that if you chop the head off a stick insect, the body will survive for some days. 'Just for a joke, and to amuse the company at the Villa', Steno and Redi put the dead head back on to the insect's body, where it stuck to the oozing wound. The apparently resuscitated insect moved slightly, excreted and even laid eggs, 'to the great surprise of all who were not in the secret'. Even here, however, Redi made sure entertainment played second fiddle to science – he continued: 'an overhasty writer would have had many eyewitnesses to vouch for the truth of this experiment, but in asserting the restoration of the heads as genuine, he would be writing sheer nonsense'.

After all this work, Redi's conclusion was simple and profound: 'no animal of any kind is ever bred in dead flesh unless there be a previous egg-deposit'. In other words, there was no such thing as the spontaneous generation of insects from decaying matter. With a series of careful experiments, designed to test alternative explanations, Redi had not only swept away thousands of years of confusion over the generation of insects, he had also redefined the notion of scientific proof.[44] By repeating his experiments under a wide variety of conditions, he was able to state with a high degree of confidence that both his observations and his interpretations were accurate. Reading his book, a modern scientist might be perplexed by the poetry, but he or she would certainly recognise the method at work.

Despite this extremely rigorous approach to discovery, Redi's work was not flawless. On one level, his biggest mistake was as interesting as his accurate descriptions. He stumbled when it came to the difficult question of gall insects – wasps, beetles and the like that hatch out of galls (brown, woody growths that are found on many kinds of plant). Each species of plant has a specific shape of gall or set of galls (the reason why there is such a wide range of shapes is still a matter of debate[45]), and each kind of gall produces a given kind of insect. Although the adult insect will bore its way out of the gall when it is ready, leaving a tiny hole, before that the surface of the gall is completely smooth and unmarked. However, if the gall is opened, it can be seen to contain soft vegetable matter and, right in the middle, an insect larva. Redi's problem was to explain how the maggot got there. His initial assumption flowed from everything he had already found out: 'I believed, or rather suspected, that galls were originated by the fly, which, in the Spring, makes a small slit in the young twigs of the oak and hides one of her eggs in the opening: the gall arising thence.'[46] Brilliantly, Redi suggested the gall might be a kind of disease caused by the female insect, similar to the swelling produced by a wasp sting (this is not so far from the truth, as galls are in fact produced by something like a plant immune reaction, in response to the laying of an insect egg). To prove his point, Redi undertook a mammoth research programme which involved opening more than twenty thousand galls. Every time, the inside of the gall was full of decaying vegetable matter and contained a single maggot.

Having noted that galls always occur in the young parts of the tree, and that each kind of gall produces a specific kind of insect, Redi suddenly concluded, 'I have changed my opinion, and I think it probable that the generation of worms in trees does not occur fortuitously, nor does it proceed from the eggs deposited by flies.' There was no reason given for this U-turn, although it is clear that, unlike the experiments on flies, in the case of gall insects Redi had not been able to satisfy his own strict criteria for accepting a proposition as true – he had not actually seen an insect laying an egg, nor traced the path from egg to adult.

Redi tried to align this position with his previous findings by suggesting that the reason why gall insects were generated from decay rather than eggs was due to the fact that the gall (or at least the plant) was alive, and that the 'soul or principle which creates the flowers and fruits of living plants . . . is the same that produces the worms of these plants'. He then went on to use this very mystical explanation to account for maggots found in cherries and other fruits, as well as apparently spontaneous parasites such as tapeworms.[47] For Redi, therefore, important parts of the animal world could not be accounted for by the general rule he set out as a result of his experimental studies.

This change of mind is very telling: having systematically excluded all possible mystical agencies from insect generation by carrying out careful experimentation, Redi suddenly drew back from the thoroughly materialist implications of his findings and left a thin thread connecting his work with centuries of tradition. Years later, in 1693, when it had been shown that his starting point was right – gall insects hatch from eggs laid by female insects – Redi confessed to his friend Lanzoni that 'I let this passage escape from my pen almost by force'.[48] From a psychological point of view, that 'force' may have been the weight of Aristotelian tradition pulling him back from the brink. More interestingly, the problem of gall insects reveals how Redi saw his science. Despite his claim that, with Steno, he was trying to 'extract general laws from our observations',[49] this final episode shows that Redi saw a 'law' as simply a series of convergent results from individual experiments, rather than a statement of something he felt was true under all circumstances because it expressed consistent underlying forces. However, in the context of seventeenth-century knowledge, this weakness was a strength. Redi based his views not on a theory, but on his observations and his interpretation of them. If, after dissecting more than twenty thousand galls, he had no direct evidence that gall insects were not produced by spontaneous generation, there was little else he could do but say so. Furthermore, by focusing on the question of gall insects, he had openly addressed the most difficult example that seemed to prove the spontaneous generation of insects, thereby highlighting his intellectual honesty and rigour. If he had rejected (or, even worse, hidden) his findings without having superior experimental evidence, that would have proved that he was still influenced by a pre-scientific method. Despite eighteenth-century observers claiming to be 'astonished' by Redi's final verdict,[50] his approach was thoroughly materialist and empirical: he called the shots the way he saw them, in the light of the experimental evidence. Unfortunately, in this case that experimental evidence, despite its abundance, lacked quality. He had observed, but not closely enough.

When Redi's book was published in Florence, readers were struck by the rigour of the experiments – another Italian scientist, Giuseppe Ferroni, later wrote to Redi to tell him that 'the clarity of style and frankness of judgement' had almost made him want to give up studying mathematics and turn to natural history.[51] As news of Redi's findings spread round Europe, thinkers in Paris, London and Hamburg became increasingly impatient as they found it difficult to get hold of a copy of his book – it had still not arrived in London bookshops some eighteen months after publication.[52] When readers were finally able to judge for themselves, they recognised both the strengths and the weaknesses of Redi's work. In the *Philosophical Transactions* an anonymous reviewer (almost certainly Henry Oldenburg) gave a detailed summary of its contents, praising the 'curious and considerable observations' it contained. But, significantly, about half the review was devoted to the contrast between Redi's general view of generation in insects and his very different understanding of the situation in gall insects. Although the reviewer did not say whether he agreed, he thought the discrepancy sufficiently important to draw his readers' attention to it – right from the opening sentence of the review.

The Dutch astronomer Huygens was clearly not convinced that Redi had got to the bottom of the matter. In May 1669 he told Oldenburg, 'You will find Redi's book very interesting, without however finding that he has yet exhausted that question, of which the chief problem – to know whether any insects arise from corruption – is not sufficiently cleared up, although much better than has been done hitherto.'[53] John Ray was of a similar opinion, implying in July 1671 that he was not satisfied with Redi's suggestion with regard to galls that it was the 'Vegetative Soul of the Plant that yields those Excrescencies'.[54]

Despite the devastating power of Redi's book, the battle against those who believed in spontaneous generation was far from over. A few years later, Kircher replied to Redi's criticisms, implying that there was something underhand about the way Redi had done his experiments: 'I . . . undertook my investigations in genuine candour,' claimed Kircher, 'not hidden in a cabinet, but openly, in the presence of the most learned professors at the Roman College . . . they were witnesses to my procedures, especially those concerning the generation of serpents, scorpions, frogs.'[55] He then went on to complain that 'just because one or the other of them was unsuccessful when he [Redi] tried it, he concludes that Kircher was wrong'. Similar arguments were put forward in 1681 by one of Kircher's supporters, Father Filippo Buonanni, who claimed that if Redi's attempts to repeat Kircher's experiments had not succeeded, this was due to poor technique and faulty reasoning. Both these rebuttals miss their mark and show the gulf between Redi's approach, based on repeating a series of experiments with a strict logic behind them, and that of Kircher and his co-thinkers, who stuck with their interpretation of an experiment willy-nilly, rather than thinking of alternative explanations and then trying to exclude those possibilities by carrying out more experiments.

By the end of the 1660s, thinkers all over Europe had been able to read Redi's arguments. Most were extremely impressed, despite the very real problem of gall insects. And within a few years even that issue disappeared as the question was resolved by other thinkers. By the 1680s, only the most recalcitrant Aristotelians continued to oppose the idea that insects were generated from eggs laid by females that had mated with a male of the same species. However, the change in the way that people looked at the natural world was not instantaneous, nor was it necessarily permanent. In July 1678, Robert Hooke, one of the sharpest and most influential members of the Royal Society, described an experiment in which he was apparently able to generate insects from decaying plants. His interpretation of his findings harked back to Boyle's formulations of twenty years earlier, as he suggested that 'the spirit or life' of the dead plant 'had flown away in the insects' that it had generated.[56] But this was not typical of the overall tendency of thought about generation. Even those, like Redi, who believed that gall insects were a case apart, still argued that this spontaneous generation was not random but lawful – each kind of gall generated its own kind of insect. Through Redi's researches a huge step was taken towards realising that there are laws of generation. It was now widely accepted by European thinkers that like – generally – bred like. Ordinary people, quite understandably, continued to believe in spontaneous generation, trusting the evidence of their own eyes and never hearing about Redi's work.

Within a few years the problem of spontaneous generation literally entered a new dimension, with the discovery of microorganisms. Finding out exactly where these incredibly tiny creatures came from, by testing the competing hypotheses, proved far more difficult than Redi's relatively simple experiments under the Tuscan sun. Nevertheless, when the question was finally resolved, late in the nineteenth century, it would use the same kind of logical argument and experimental evidence developed by Redi to study his insects.[57]

4

THE TESTICLES OF WOMEN

When Steno arrived in Tuscany in 1666 he carried a letter of recommendation from Thévenot addressed to 'friends and patrons' in Italy.[1] This letter, together with his reputation, gave him access to Grand Duke Ferdinando's court, which was in its winter quarters at Pisa. Ferdinando welcomed the visitor, and by the summer the twenty-eight-year-old Steno was an important contributor to the buzz of science that had become an essential component of court life in Tuscany. As well as helping Redi with his experiments on the generation of insects, Steno was also able to show off the dissecting skills that had so impressed Paris the year before. Ferdinando was delighted to have such a talent in his court, and gave the young man everything he needed: bed, board and all the facilities he required. As Steno put it: 'He only has to nod, and all that is necessary for my scientific researches in whatever field I choose, is put at my disposal. For my anatomical studies there are various animals; then, too there are the cadavers at the hospital.'[2] In return he simply had to be an entertaining court scientist.

Less than six months after arriving in Florence, Steno produced his first Tuscan manuscript, a sixty-eight-page study of the function of the muscles, *Elementorum Myologiae Specimen* ('A model of elements of myology' – the study of muscles). Bound together with two other studies, it finally appeared in March 1667, a few months before the publication of the *Saggi* (the accounts of the experiments carried out by the Accademia del Cimento) and printed at the same Florentine print-shop, Stella ('Star'). After Redi's 1664 study of the viper, Steno's book, with its gushing dedication to Ferdinando, was the first public sign of the work being done at the Tuscan court. Its impact was so great that the Medici brothers, Ferdinando and Leopold, undoubtedly got their money's worth in terms of prestige and the strengthening of their reputations as patrons of science.

Steno's starting point was his previous anatomical work on the structure of muscles, which he had carried out in Leiden and Copenhagen. Going beyond his descriptions of muscle fibres, he now tried to understand how muscles functioned, using geometrical arguments backed up with further anatomical facts contained in a letter to Thévenot, which he reproduced at the end of his study. As he put it: 'I wished to demonstrate in this dissertation that unless Myology becomes part of Mathematics, the parts of muscles cannot be distinctly designated nor can their movement be successfully studied.'[3] This objective was much more important than the detail of what he discovered (his findings were rejected by his contemporaries, and his work was overlooked for centuries before scientists recently realised that he was basically right).[4] A key ambition of the scientific revolution was to provide numerical, objective descriptions of all aspects of the Universe, including living and anatomical phenomena. Steno adopted this approach after coming into contact with the members of the Accademia del Cimento, whose concerns were more systematically mathematical than those of his teachers and colleagues in Leiden, Copenhagen or Paris – none of his previous studies of anatomy or natural history had shown any sign of this way of looking at the world.[5]

By March 1667, long after the *Elementorum Myologiae Specimen* had been approved by the censor – the Holy Office required that books should contain nothing that was 'against Catholic Faith or good morals' – Steno had added two further sections to his manuscript, both of which would be extremely influential. Neither of the new pieces had any connection with the material on muscles, and both had the same apparently unpromising starting point – the dissection of a shark.

Towards the end of October 1666, French fishermen off the Tuscan port of Livorno spied a gigantic great white shark (*Carcharodon carcharias* – the kind of monster portrayed in the film *Jaws*), close to the surface of the sea. Brave or foolish, they managed to catch the terrifying fish and winch it ashore, where the huge thrashing animal – it weighed around 1200kg – was tied to a tree and clubbed to death. After cutting out its liver (claimed to weigh 100kg), the fishermen hacked off its head and then rolled the rest of the massive carcass back into the sea. News of the catch soon reached Florence, and Ferdinando immediately ordered the shark's head to be brought to the Tuscan capital for Steno to dissect in front of select members of the court.

Steno's investigation of the shark's head was described in a forty-page account entitled *Canis Carchariae Dissectum Caput* ('A Carcharodon-head dissected') which formed the second section of *Elementorum Myologiae Specimen*.[6] Like a true courtier, Steno presented his study as a kind of amusement for Grand Duke Ferdinando. The opening sentences, which directly followed the article on muscles, prefigured Redi's sophisticated self-deprecation in the pages of *Esperienze intorno alla generazione degl'insetti* (published the following year): 'I have no doubt that a long and uninterrupted account of observations of muscles would be distasteful to the reader; accordingly, since variety is the spice of life, I have decided to add to what has gone before material which should provide an opportunity to review various isolated observations.'[7] In other words, Steno's presentation was designed to entertain his patron.

The truly innovative part of this second section of Steno's book had nothing to do with its main topic, which was anatomy. Towards the end of his account, he highlighted the similarity between the teeth he observed in the Livorno shark and the *glossopetrae* ('tongue-stones') which were found on exposed ground throughout southern

Europe. These stones were commonly thought to be vipers' tongues (although they looked nothing like them), but were in fact fossilised sharks' teeth. Steno, like a number of previous thinkers, suggested that *glossopetrae* looked like sharks' teeth because that is what they were. This attractively simple idea had the enormous disadvantage of requiring an explanation of how sharks' teeth got deep inland, even to the tops of mountains. Here Steno went further than anyone else and came up with a series of explanations as to how the remains of various kinds of sea creature, including *glossopetrae*, could be found in the ground, sometimes far from any open stretch of water.

Like a good Christian, Steno took as historical truth the two biblical accounts of times when the Earth was covered with water – at the Creation and during the Flood. On both occasions, sharks and other marine organisms could have been stranded by the receding floodwaters, he said. More interestingly, he also pointed out that earthquakes could lead to massive changes in the surface of the Earth: part of the sea bed might be thrown upwards, becoming dry land and bringing a host of animals with it. In a few short sentences, Steno outlined a basically correct explanation of the formation of sedimentary rocks and even of the fossilisation of organic matter. He did not, of course, have any notion of the immense depth of geological time – although he never wrote on the matter, he would have held some variant of the contemporary opinion that the Earth was around six thousand years old. This view, which was based on adding up the 'begats' in the Old Testament, was challenged around a century later by the Scottish geologist James Hutton, and was finally disproved to the satisfaction of everyone except hard-line fundamentalists at the beginning of the twentieth century.

Steno was well aware that his explanation of fossils, which went against the ideas of Aristotle and the ancients, might be problematic for the Church and thus embarrassing for his royal patron. So he used Galileo's device of claiming that the view he had outlined was merely one possibility amongst many: 'While I show that my opinion has the semblance of truth, I do not maintain that holders of contrary views are wrong. The same phenomenon can be explained in many ways; indeed Nature in her operations achieves the same end in various ways. Thus it would be imprudent to recognise only one method out of them all as true and condemn all the rest as erroneous.'[8] In other words, fossils might be traces of long-dead organisms or, then again, they might have been put there by God. Steno was hedging his bets.

Over the next few months Steno developed his idea as he and Redi walked around the Tuscan countryside, trying to understand the geological formations they observed. Two years later, in a *prodromus* ('advance notice') to a document about fossils to be wittily entitled 'Dissertation on solids naturally contained within solids', Steno gave a full account of his conception of geological change, including his decisive suggestion that geological layers could be 'read back' as a record of the past. His principle of superposition, as it is now called, marked the beginning of geology.[9]

Having found a mathematical explanation of one of the most widespread living phenomena – movement – in the first part of *Elementorum Myologiae Specimen*, and then laid the bases for the scientific study of the Earth and its history in the middle section (*Canis Carchariae Dissectum Caput*), the final part of Steno's book, entitled *Historia Dissecti Piscis ex Canum Genere* ('Study of the dissection of a dogfish'), looked as if it would be a disappointment. It was shorter than the other two parts – a mere nine pages long – and lacked both the mathematical precision of the material on muscles and the drama of a giant shark with a mouth full of vicious teeth. In this brief account, Steno described the dissection of a small female dogfish that gives birth to live offspring, which was neither rare, nor terrifying, nor particularly mysterious.

After discussing the anatomy of the head – like its massive relative, the dogfish has a tiny brain, Steno noted – he moved to the other end of the animal, where he found a number of small fish in the stomach, which had apparently been swallowed whole. Although it was pretty tame stuff, this was novel; so few animals had been studied anatomically that any competent study was bound to come up with something hitherto unknown. But in the last couple of pages

Steno used a simple analogy and, in a few lines, made a huge break-through in humanity's understanding of generation.

First, he described the similarities between the reproductive tract of the viviparous dogfish and that of the egg-laying ray, which he had dissected several years earlier. Primed by his previous work on generation with Thévenot, and by the dissections of the female genital organs in women and sheep which he had carried out in Leiden at the beginning of the decade,[10] Steno then began thinking about the nature of generation in oviparous and viviparous animals, and came to the following amazing conclusion: 'having seen that the testicles of viviparous animals contain eggs, and having noticed that their uterus opened into the abdomen like an oviduct, I have no doubt that the testicles of women are analogous to the ovary, whatever the manner the eggs themselves, or the matter that they contain, pass from the testicles to the uterus'.[11]

And that was that. No further dissections, no experimentation, no detailed drawings – merely the declaration that he had seen eggs in the female 'testicle' (the ovary), and a crude sketch showing the parallels between the female reproductive tracts of viviparous and oviparous fish. Steno had made his point and then passed to his closing remarks about sharks. 'The testicles of women are analogous to the ovary': this simple statement was as powerful as the '*Ex ovo omnia*' that appeared on the frontispiece of Harvey's *De Generatione*, yet more far-reaching, for the simple reason that it was more precise. Harvey may have thought that humans came from 'eggs', but he was completely unclear about exactly what that 'egg' was. Steno not only said that he thought human eggs were like the eggs of other animals; even better, he told his readers where they could be found: in women's 'testicles'. In the final part of his description, however, he hinted that there might be some difficulty in understanding how eggs – 'or the matter that they contain' – got from the ovaries to the uterus. With this, Steno had shrewdly put his finger on a major problem that would cause supporters of the egg theory of human generation no end of difficulties in the following years.

Although Steno's description of his egg hypothesis was typically low-key, a mixture of his natural modesty and the courtly rhetoric that predominated in Florence, it contrasted sharply with the deliberate ambiguity of his view of the origins of *glossopetrae*. Whereas he had carefully suggested that *glossopetrae* might have been formed in many different ways, when it came to the nature of the 'testicles of women', he was categorical: 'I have no doubt.' This may imply that he thought his theory of generation was less potentially explosive than his ideas about fossils, or it may have simply reflected his greater certainty in the case of the ovary, born out of his confidence in his dissections and a sense that animals tend to have very similar internal structures, no matter how different their external appearance.

With his simple suggestion, Steno had exposed the inadequacy of the prevailing idea that the female 'testicles' were equivalents of the male organs that were either degenerate (as Aristotle argued) or that produced a thin female 'semen' (as proposed by Galen and by some popular anatomists such as Thomas Raynalde). In so doing he had implied there was a common basis to the generation of all animals, both viviparous and oviparous, vertebrate and invertebrate: all female animals, including women, had ovaries, and within those ovaries were eggs. This view of human generation was radically different from the views that had predominated for two thousand years: it identified the woman's contribution not as 'semen', nor as 'menstrual blood', but as an egg.

However, Steno would have been horrified had he realised the profound implications of his idea. By suggesting that human generation is no different from that of any other animal, he enabled science to take a tiny step on the road to a materialist understanding of the ultimate origins of all life. As this understanding grew over the centuries, the role of Steno's God was conversely reduced until He was eventually squeezed out of scientific thinking altogether, discarded as an unnecessary hypothesis, a superstitious relic of humanity's ignorance. Steno, like the Church censor who literally gave his stamp of approval to his book, had no inkling where this idea would eventually lead.

Steno's hypothesis had an immediate, far less dangerous, impact. In its simplest, most literal form, it was taken up throughout Europe, as many thinkers promptly accepted that Steno was right, despite the fact that his argument was based entirely on analogy, without a shred of anatomical or experimental evidence apart from a brief, unsubstantiated and vague claim to have observed the eggs. Steno's readers were well aware of the limits of his demonstration, and sensed that more was to come: an anonymous reviewer of his book indicated that 'further proof' of his hypothesis would be provided in the future.[12] But even without such evidence, the idea was so attractive, so obvious, that it met with widespread support.

The transformation of scientific ideas about the nature of the female organs of generation was so rapid and so complete that within a few years there was virtually no hint that modern thinkers had ever seen things differently. In 1679, the *Journal des Sçavans* crowed: 'The view that man, as well as all other animals, are formed from eggs, is something that is now so widespread that there are hardly any new philosophers who do not now accept it.'[13] However, some writers seemed to have a flimsy grasp of what was actually involved. In 1674 the Frenchman Louys Barles wrote a popular account of the recent discoveries of human generation in which he described the growing embryo and claimed that 'after a certain time destined by Nature, their shell cracks, their membranes are broken, and we have the pleasure to see a man come out of an egg'.[14] Just as it took some time for this literal view of 'eggs' to disappear, so the old ideas about the ovaries lingered on in the language. Even in 1673, Steno was still using the term 'testicle' to describe the mammalian ovary when he reported the dissection of a female deer, although he used 'ovary' to describe the situation in vipers and fish.[15] By the second half of the 1670s, however, thinkers were happily writing about 'women's ovaries' – with not a cracked eggshell in sight.[16]

Many people accepted Steno's hypothesis simply because it made sense, and fitted in with the alchemical and scientific attempts to find the material bases of generation that had dotted the century. In the case of Jan Swammerdam, his knowledge of his friend's integrity undoubtedly reinforced his confidence in Steno's suggestion. But Swammerdam also had another reason to accept the idea without question. He had already seen what he thought were human eggs, over a year before Steno had published his hypothesis.

At the beginning of 1667, at the same time as Steno was finalising the two sections of his book that dealt with sharks, Swammerdam was back in Leiden, putting the finishing touches to his doctoral dissertation on respiration, getting it printed and rehearsing for his examination, which involved a difficult public question-and-answer session. There was also a ceremonial procession to endure, as students and professors trooped around Leiden in full academic regalia. In the middle of these preparations, Swammerdam received a message from his teacher of anatomy, Professor van Horne, asking him to come to his home and help with the dissection of a female cadaver. The very fact that the body was to be studied in the professor's house rather than in the university amphitheatre showed that Van Horne's intentions focused on research rather than teaching. This was not to be a normal public display, but an attempt to find something new, requiring discussion and concentration. An audience would hinder, not help.

Van Horne's aim was to study the female organs of generation and provide a clear description of their structure and organisation. This was a continuation of his unpublished 1665 study of the male testicle, in which he had tried to understand the different fluids, or semen, that it seemed to produce (Van Horne rightly argued that the male ejaculate appeared to be composed of substances with different consistencies, each of which could be traced to a particular part of the testicle). Van Horne apparently needed the aid of his former pupil more for his artistic ability than for his skill with the scalpel. The older man was no slouch when it came to dissection, and required little help in that respect. But he not only needed to find out how the female reproductive system was organised; above all he had to convince other people of what he had seen. Knowing whether to believe what someone said was a major problem for the scientific revolution, especially given that so many fantastic and hitherto unsuspected things were being discovered. One widely used method involved invoking illustrious (and preferably aristocratic) observers who had seen the particular experiment or demonstration. This form of seventeenth-century scientific name-dropping implied that these various celebrities could vouchsafe for the truth and accuracy of the account, thereby giving it more weight. One variant of this was communal 'witnessing', when a group of scientists, such as the Royal Society or the Accademia del Cimento, published an experiment under their collective name.

An even more convincing approach was to use illustrations; this was useful where the subject lent itself to pictorial representation, such as dissection, but less appropriate when it came to describing the outcome of, say, experiments on motion. Scientific illustrations were particularly popular following the introduction of copper-plate engraving, which became widespread from the middle of the seventeenth century and which allowed illustrators to present very detailed images. Then as now, scientific illustrations had multiple levels of meaning. As well as demonstrating what had been seen, illustrations helped the observer to see more clearly what he had discovered, refining his understanding. In other words, as well as showing what was observed, they were also an integral part of the study, helping to shape the viewer's vision. And by acting as a guide for subsequent investigators, framing what they expected to find, the 'proof' provided by an illustration also furnished the basis for testing the discovery by trying to replicate it. Finally, illustrations valorised the research by presenting striking or beautiful images that both transmitted the scientists' aesthetic judgements and reflected their skill either directly (in the case of those who made their own drawings) or indirectly (if someone else intervened).[17]

Swammerdam was a highly skilled draughtsman who during his career made hundreds of drawings for his own studies, mainly showing the results of dissections (some on incredibly small insects), but also including striking illustrations of whole animals. Many of his original drawings still exist and are remarkable for their detail and the flowing movement of the pen or brush; Swammerdam was not

simply an expert scientific illustrator, he was also an artist whose subject was anatomy. His outstanding ability to draw what he saw would
have been enough to get Van Horne to ask for his help. But
Swammerdam could do more than simply *draw* dissections: he could
make them visible, permanently, by injecting hot coloured wax into
actual blood vessels. The result was stunning. The blood vessels were
filled out with the red, yellow or green wax, making them shine and
bulge; even the finest capillaries could be traced, showing their precise course and interconnections. Information that a drawing took
hours to provide could be obtained in a matter of minutes, in
unprecedented detail. Furthermore, if the material was preserved in
alcohol or some similar substance, the coloured preparation could be
kept for decades or more as both a study aid and a permanent
demonstration of the discovery.[18]

Swammerdam was not the only person to use injections as a tool
for the investigation and presentation of anatomy.[19] In 1663 Robert
Boyle had proposed using a number of substances including plaster
and gelatine, while both Malpighi and De Graaf had gone a step further and employed various coloured inks in their studies of the
kidney and the male genitalia, respectively. Another of Swammerdam's
student friends, Frederick Ruysch, would soon surpass all his contemporaries in his ability to preserve and present bodies, often in a
distinctly macabre fashion, with babies preserved in their baptismal
clothes as though they were asleep, and embryonic skeletons playing
minute violins.[20] But at the beginning of 1667, Swammerdam was
still the best, as shown by the enthusiastic reception of a liver he had
prepared in this way, which was put on show in the Leiden
University medical amphitheatre.[21] That was why Van Horne
wanted his help.

Van Horne's interest was initially focused on the human uterus
and the Fallopian tubes, which, he pointed out, shared fundamental
similarities with egg-laying animals such as frogs, birds, lizards and
'aquatic salamanders'.[22] As the dissection proceeded, he suggested
that the female 'testicles' looked like the ovaries in egg-laying animals –
exactly the same insight achieved by Steno. Swammerdam and Van

Horne dissected these structures and inside found vesicles – small fluid-filled bubbles – which they decided were, or contained, eggs. The eggs somehow found their way to the uterus through the Fallopian tubes, as in egg-laying animals where the egg passes through the oviduct. They even found some evidence that these vesicles contained eggs when they dissected a cow: when the vesicles were boiled, they turned white, just like a chicken's egg.

At Van Horne's request, Swammerdam made some preliminary sketches of the dissection to go with the coloured wax injections, before turning back to his doctoral dissertation. Once he had finished his exams in spring 1667, he returned to his father's home in Amsterdam, where he continued his studies of insects. The idea was that Van Horne would soon complete his investigations, then the whole thing would be published, together with Swammerdam's drawings. But Van Horne did not keep to the plan. Four months after the dissection, he wrote to Swammerdam, promising that his final manuscript would soon be in the post. It was never sent. As the year wore on, Van Horne promised first to finish his draft 'after the holidays are over', but then sent Swammerdam an embryo to dissect and study. For the rest of the year, nothing more was heard of Van Horne's investigations, nor of his insight that women's 'testicles' were in fact ovaries. Then, at the end of 1667, he insisted that Swammerdam and his friend, the Amsterdam physician Matthew Slade, come to Leiden and help him dissect 'a delirious virgin who had drowned herself'.[23]

Swammerdam responded to this call, but while he was in Leiden, at the beginning of 1668, matters suddenly became more urgent. Despite his close friendship with Steno, he knew nothing of Steno's ideas about eggs, nor had he seen Steno's latest publication. The circulation of books was often incredibly slow, and sometimes did not even involve commercial channels: when Steno's *Elementorum Myologiae Specimen* and its bombshell about the female organs of generation finally appeared in northern Europe, it was not on the shelves of a bookshop, but in the baggage of a lovelorn prince.

✳

In 1661, Prince Cosimo, the only son of Grand Duke Ferdinando of Tuscany, had married Marguerite Louise d'Orléans, a cousin of Louis XIV, when he was nineteen and she was still only fifteen.[24] This was not a marriage of love, but an alliance that welded the interests of Tuscany and France. Marguerite was a strong-willed young woman who did not appreciate being forced to marry a man she had never met, nor having to abandon her French lover. On arriving in Florence, she made these feelings known to Cosimo, and indeed to the whole of the Tuscan court. Despite the birth of two children, in 1663 and 1667, relations between Marguerite and Cosimo were difficult. Cosimo's strict religious upbringing had left him a sad young man, and he grew even more melancholy as his marriage became increasingly strained. Caught between an oppressive, pious mother and a determined, irritated wife, Cosimo was not happy.

In autumn 1667, following a furious row between Cosimo and Marguerite, Ferdinando decided that if his son became a bit less dour and a bit more sophisticated, Marguerite might be nicer to him and further grandchildren might come along. So Cosimo, now twenty-five, was sent to visit northern Europe, in particular England and the Dutch Republic, accompanied by Magalotti, the one-time Secretary of the Accademia del Cimento. Ferdinando's strategy eventually worked, but it took two such journeys, in 1667–8 and again in 1668–9, before Marguerite could stomach Cosimo's presence long enough for the necessary to occur. A second son, Gian Gastone, was finally born in 1671, a year after Ferdinando had died and Cosimo had become Grand Duke. Gian Gastone was the last male Medici.

On Monday 9 January 1668, on the first of his trips, Cosimo visited the Leiden medical school.[25] During the royal inspection, one of his aides gave Van Horne a copy of Steno's *Elementorum Myologiae Specimen* – Cosimo had brought several copies with him, to give away as tokens of esteem and as signs of the work being done at the Tuscan court. Swammerdam saw the book, and immediately realised the importance of the section dealing with the reproductive system of the shark: while Van Horne had dithered, Steno had published.

Even at its beginnings, science was intensely competitive. One of the main ways competition expressed itself was that credit for a given discovery went to those who published first. Being second got you nowhere. This is still very much the case, and the reasons that lie behind it are pretty similar. The explanation is not so much that most scientists are men and therefore tend to be egotistical and aggressive, although both are true. The sad reality is that scientists compete over prestige and acclaim because, most of the time, these are the only things that are at stake. Few scientists make a fortune from their find-ings, and the recognition of their peers is all they can really hope for. And that recognition goes to the person who makes a discovery first, or who is able to demonstrate a principle in the most novel and strik-ing way. Nowadays such recognition and prestige can lead to professional advancement and thus financial and employment advan-tage. This was far less the case in the seventeenth century, when there was no such thing as a scientific profession, and most 'scientists' were either amateurs like Swammerdam and Steno, or academics like Van Horne whose job was to teach, not do research.

The importance of recognition as a motivation for scientific research highlights the fact that science is an intensely social activity. Having a brilliant idea, or making an earth-shattering discovery, is pointless if no one else knows about it. Science has to be communicated, to enter the social sphere, before it means anything. Swammerdam's reaction when he read Steno's book shows that he was acutely aware of both these aspects. Van Horne might have been the first to realise that the female 'testicles' were in fact ovaries, but as far as the rest of the world was concerned, Steno had priority. Whatever prestige might accrue to the discoverer of women's eggs – and there would surely be a great deal – none of it would come the way of Van Horne and Swammerdam, no matter how justified their claim.

Shocked and worried, Swammerdam immediately went to see Van Horne and upbraided him for his 'tardiness in publishing'. Having relieved his irritation and frustration, Swammerdam then wrote a letter to Steno, explaining that Van Horne had suggested the female 'testicles' were ovaries before Steno had published, and hoping that Steno would not be annoyed. Within a few weeks, in March 1668, Steno replied with a grace that spoke volumes for his friendship with Swammerdam, his respect for Van Horne and his good nature: 'Dearest friend: I await with great expectation the observations of yourself and the most distinguished Doctor Van Horne on the testicle, and I am far from aggrieved that I have been preceded in this matter by my Teacher and my friend, and I will solemnly swear that I will declare that had I known it, I would not only have mentioned him by name, but would have made public his observations in the same place as my own.'[26] This letter remained private until 1672, when it was published by Swammerdam as part of his growing conflict with De Graaf.

One additional reason for Steno's generosity might have been his recent conversion to the particularly pious strain of Catholicism that seemed to predominate in Tuscany.[27] In summer 1666 he had watched an enthusiastic crowd in Livorno gathered round the 'host' (the piece of bread which Catholics claim has been transmuted into the body of Christ). Steno was struck by the fervour of the participants and began to doubt his Protestant convictions, which were further undermined by a series of long theological discussions with Lavinia Cenami Arnolfini, wife of the ambassador of the city state of Lucca and an extremely devout Catholic. Then, in November 1667, as he was walking down a street in Florence trying to find a friend's house, Steno heard a voice calling from a window telling him to 'go on the other side' in order to get to his destination. His state of mind was obviously quite fragile, for he took this simple direction as a divine instruction, and decided on the spot to convert.

For a few years Steno pursued both his scientific work and his inner spiritual journey. As well as his work on fossils – 'solids contained within solids' – he continued to investigate generation, carrying out a series of dissections of bears, tortoises, fish, vipers, deer, wolves, donkeys, pigs and mules.[28] His study of mules was particularly striking. Mules – the offspring of a female horse and a male donkey – are virtually always sterile. Steno noted that in one mule he

dissected there were no 'eggs', while in another he could see only a
few. He concluded that a female mule might occasionally be able to
reproduce, but that the absence of 'eggs' was probably not the reason
why mules are infertile. This was absolutely right – hybrid sterility is
a complex phenomenon, and the absence of eggs is not the main
reason for it. As one expert in mammalian reproductive biology
recently put it, discussing Steno's work on the mule: 'We have been
unable to improve upon this conclusion, even three hundred years
later.'[29]

Eventually, Steno found it impossible to continue being a scientist
and a priest (he had been consecrated in 1675), and in 1677 he abandoned science altogether, becoming a bishop and leaving Italy to
work amongst the Protestant unbelievers in Germany. Although he
maintained contact with many of his scientific friends – including
Leibniz, who tried to win him back to science – he remained a convinced Catholic until the end. In 1686 Steno died in poverty in
Germany; his body was returned to Italy and was eventually buried
in the Medici family church in Florence, the Basilica di San Lorenzo.
Three centuries later, in 1988, he was beatified by Pope John-Paul II.

The first part of 1668 was a bad time for Van Horne. At the beginning of the year he had learnt that Steno had beaten him into print
over the true nature of women's 'testicles'. Then, a few weeks later,
in March, he got wind of the imminent publication of a book by
another of his one-time students, Reinier de Graaf, on the other
important aspect of generation that Van Horne had been studying for
several years: the structure of the male genital organs. This collision
of interests led to a crisis between teacher and student and a very
public polemic.

After his visit to France to obtain his doctorate, De Graaf had
returned to the Dutch Republic in 1666 and had settled in Delft, a
beautiful town of around twenty-five thousand people, not far from
Leiden, where he practised medicine. Although his practice proved
extremely time-consuming, De Graaf managed to maintain both his
research and his particular brand of investigative surgery and

medicine. He returned to France briefly the next year, where he again discussed his work with Montmor, including his studies of the male organs of generation. By the beginning of 1668, still only twenty-six, De Graaf was back in Delft for good.

De Graaf's decision to settle in Delft was apparently driven by financial pressures. He needed somewhere to establish a medical practice (unlike Steno and Swammerdam he could not count on rich patrons or help from his parents). Although Delft was neither an academic centre like Leiden nor a major political city like Amsterdam, it did have a substantial potential patient base in the shape of a sizeable middle class. Delft had a large pottery industry; the development of 'Delftware' – earthenware pottery glazed in white and decorated in blue, which in the seventeenth century was produced by some thirty factories in the Delft region – spurred the town's economic growth. The money made by the factory-owners spilled over into the rest of the city, creating a comfortably well-off layer who could both afford to pay their physician and to buy works of art created by the members of the Delft artists' guild. By the 1660s, the number of artists had declined, as many left for Amsterdam or Leiden. One of those who remained was Johannes Vermeer. When De Graaf arrived in Delft, Vermeer had just completed a spurt of amazing creativity, producing several masterpieces (including *Girl with the Pearl Earring*) and culminating in what he apparently considered his greatest work, *The Art of Painting* (1667). De Graaf may have left Paris behind him, but Delft, despite its smaller scale, could rival the French capital through its intensely creative atmosphere where art, science and commerce flourished side by side.

After his return to the Netherlands, De Graaf heard about Van Horne's study of the testicle. This was not a secret: throughout 1667, Van Horne had discussed his work with many people, while Swammerdam, who had tried some of his wax injection techniques on the testicle, talked about the work with De Graaf. Worried about being upstaged by his former professor, on 20 February 1668 De Graaf wrote a *prodromus* of his book in the form of a letter addressed to Sylvius, the man who had been his closest teacher and who was

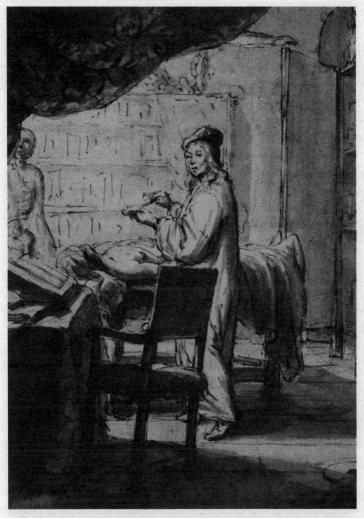

Reinier de Graaf dissecting in his rooms in Delft. The portrait, attributed to Jan Verkolje, may have been a sketch for an oil portrait that was either never completed or has since been lost. For a discussion of the attribution of this illustration, see Cetto (1958).

still Van Horne's colleague.[30] In this letter, which he printed in order to stake his claim to priority, De Graaf rapidly summarised his findings in the broadest terms, dealing with both his anatomical work on the structure of the male genitalia and his invention of a self-administered enema syringe, which he prescribed for a wide variety of ailments.

At the beginning of March, Van Horne got a copy of De Graaf's *prodromus* the day it came off the presses at Leiden, and was aghast. He was in danger of being upstaged by a former student for the second time in a few weeks. To protect his intellectual property, he had to act. First he tried to get letters from Swammerdam and Matthew Slade, who had helped dissect the 'delirious virgin' the previous year, in order to show that he had been working on the male genitalia for nearly three years. But Swammerdam was convalescing in the countryside, while Slade, perhaps sensing that this was a potentially delicate issue, used his friend's absence as a reason not to reply. With no one to back him up, Van Horne had no choice but to publish his own *prodromus*, in the shape of a letter to his old friend Werner Rolfink, Professor of Anatomy at Jena University.[31]

In his 1500-word document, Van Horne summarised his findings on the structure of the male organs, and stated that he had carried out his investigation before De Graaf, and without any knowledge of his former student's work. Killing two birds with one stone, he also stated his claim on the question of the female organs of generation, declaring, like Steno (although he did not mention him by name), that 'the testes in women are what the ovary is in the ovipara, namely they contain perfect eggs'.[32] Van Horne, however, went further than his ex-student. Although Steno said he had seen the eggs and that they were to be found in the ovary, he had not been more precise. Van Horne, though, closed his *prodromus* by specifying exactly what the eggs looked like: they were 'full of liquor, and encompassed with a skin of their own'. This clearly implied that the pea-sized ovarian vesicles were the eggs (in fact they contain the tiny single-celled egg). Despite the fact that this claim was made with no supporting evidence, it was picked up by the Royal Society's reviewer (almost certainly George Ent), who published a brief summary of both De Graaf's and Van Horne's advance notices, almost as soon as they appeared.[33] The reviewer, unlike Van Horne, also pointed out that Steno had previously said basically the same thing.

To explain his failure to publish any illustrations to prove his point, Van Horne blamed the 'laziness of the engraver' – which implied that the drawings existed. Furthermore, he continued, 'those who wish to see the truth of what I describe can come to my house, for I keep all the parts well prepared, through which the visitor can satisfy their curiosity', suggesting that Swammerdam's wax injection skills had been well employed and that a preserved uterus was now in Van Horne's hands. Finally, Van Horne unnecessarily raised the stakes in terms of De Graaf's claims over the male genitalia. After accepting that two pieces of work could look so similar as to give the impression that the same person was responsible for both, he then continued: 'lest the door be opened to lawsuits and insults, I will not make false accusations'. In other words, he was hinting that De Graaf had stolen his discovery.

Two months later, De Graaf responded in kind, explaining why he had published his *prodromus*:

> As a result of the excessive liberality with which I communicated and divulged my findings concerning the genital parts, certain men, who always want to glory in the findings of others, were beginning to ascribe to themselves the glory of what I had discovered. For this reason and because I had not yet arranged my findings in good order and had, moreover, calculated that a lot of time was needed to engrave the figures, I published a letter addressed to the distinguished gentleman Frans de le Boë [Sylvius] as a kind of advance notice of my findings in order to curb the insolence of these spiteful fellows.[34]

Although he did not name the 'spiteful fellows', Van Horne was clearly the target. De Graaf went on to jeer at his former teacher's *prodromus*, remarking pointedly that it was published only after his own letter, and contained 'findings about which he had preserved until then a most profound silence'. Finally, he challenged Van Horne to 'point out my errors, if any, and their causes, and in this way teach posterity that truth is to be sought through argument and demonstration and not through brawls or calumnies'.

This was strong stuff. De Graaf was impugning his former teacher's integrity, accusing him of plagiarism and of lying. Sylvius, to whom De Graaf's original *prodromus* was addressed, apparently encouraged his former student's attitude; in a letter to De Graaf, sent in March 1668, he wrote: 'the world is full of ignorant and jealous people who will not hesitate to steal other people's discoveries'.[35] De Graaf's behaviour was the first public sign of his tendency to react with anger and suspicion when dealing with the claims of apparent rivals. Over the next five years this would lead to some significant clashes in which his image of a cool-headed, brilliant young man would be brought into question. Although claims about plagiarism or outright theft were not uncommon at the time – Steno had attacked Gerard Blaes over the parotid gland duct, while Swammerdam had suggested (quite mildly and apparently incorrectly) that Ruysch had copied his drawings of the lymph vessels[36] – De Graaf's tone was unusually aggressive.

It might be expected that Van Horne had learnt his lesson, and that in the months that followed he would publish something. Anything. But he did nothing. He did not even reply to De Graaf's detailed criticisms when these were published later in the year. Instead he exchanged a series of letters with Swammerdam, who for most of this time was either ill or preoccupied with his work on insects. Swammerdam provided him with rough sketches of their dissections of both the male and female genitalia, and in the summer of 1668 Van Horne eventually returned them, marked up and ready for Swammerdam to make a fair copy so that they could be sent to the engraver. But Swammerdam either would not or could not do this, and Van Horne never received the final copies. In December he predicted it would be another year before his book was published, because of delays in making the engravings.[37] At the beginning of 1669 things were held up again, this time by Van Horne himself, as he now had access first to a male cadaver, then to a female, and wanted time to dissect them and then study his findings. From an historical point of view, his dithering was his undoing.

From August 1669 to January 1670, an outbreak of an unknown but highly contagious disease paralysed life in Leiden. Around forty thousand people were taken ill, and two to three hundred died each week. The university was closed and the teaching staff were severely affected. Sylvius lost his son to the disease and fell gravely ill himself. Overall, seven of Leiden's fifteen professors died during this period, five of them from the epidemic.[38] One of the victims was Van Horne, who died in January 1670 aged only forty-eight, without publishing anything of substance on what should have been the summit of his career: the discovery of human eggs. 'Publish or perish' goes a modern academic saying. In Van Horne's case, this was cruelly correct.

While Van Horne procrastinated, De Graaf published. In May 1668 the young man completed his study of the male genitalia and sent it to the printers, together with accounts of the use of enemas and of anatomical injections. De Graaf was aware that his main subject was tricky territory and that the 'disrespectful and lascivious' might 'try and seize upon what I have made public concerning the genital parts for lewd imaginings and scurrilous jokes'. His defence was that he had set out his findings as genteelly as possible, and that 'no one therefore can be offended by them in the slightest unless he wants to be'.[39] De Graaf's book, with the snappy title *De Vivorum Organis Generationi Inservientibus, de Clysteribus et de usu Siphonis in Anatomia* ('Treatise concerning the generative organs of men; on enemas and on the use of syringes in anatomy'),[40] contained a summary account of the anatomy of the male genitalia, with repeated criticisms of the positions outlined by Van Horne in his *prodromus*, together with sections outlining De Graaf's pet therapy (the enema) and his ideas about injecting ink into cadavers in order to understand the distribution of the blood vessels.

Most of the disagreements between De Graaf and Van Horne over the male genitalia related to points of anatomical detail, with no great consequence for the overall question of generation. Even on an issue where it might appear that there was an important difference between the two men – De Graaf insisted that there was only one kind of semen, whereas Van Horne argued there were three kinds – on closer inspection the dispute became less clear. De Graaf did not disagree that there were different components within the semen (indeed, he identified the origin of the prostatic fluid which makes up an important part of the male ejaculate and comes from the prostate gland, not the testicles); he simply thought that all the parts constituted a unitary substance.

De Graaf's description of his use of injections in anatomy might have been expected to provoke Swammerdam, who had developed the technique to a high art. But De Graaf's presentation was relatively low-key, and his injection of coloured inks was much less effective than Swammerdam's hot-wax method. De Graaf's procedure was flawed by the fact that it was temporary; veins and arteries could be easily highlighted – he claimed it was possible to reveal the whole circulatory system in a matter of minutes – but the staining would soon seep away and fade. It was useful as an aid to description, but, unlike Swammerdam's technique, it could not be used to convince people who were not present at the actual dissection.

Although the uninitiated reader can find the anatomical detail a bit dull, De Graaf's book is fascinating, opening with a lively historical and cultural summary of views of the structure and function of the male genitalia, including some unlikely accounts of Dutchmen with three testicles. This mixture of culture, anatomy and pathology was typical of De Graaf's style – an attractive blend of learning, novel findings and practical information. His discussion of the mechanisms underlying penile erection, which he put down to the swelling of the nerves produced by the passage of the 'animal spirits' and the influx of blood into the spongy structure of the organ, was enlivened by a description of how to make a dissected penis erect by filling it with water, thereby suggesting hours of puerile fun for future generations of medical students.

Some of the material in De Graaf's book has a surprisingly

modern ring. Right from the outset, he opposed what he called Aristotle's 'insulting' view of women. The Greek philosopher described a woman as 'an incomplete male', whereas De Graaf argued that 'Nature had her mind on the job when generating the female as well as when generating the male, for without females there would be no generation of any animal. And indeed being female is a kind of perfection of individual form, because sex is distinct only in perfect individual forms.'[41]

This refreshing attitude to women extended to his emphasis on the mutually pleasurable side of sex: 'Nature decided that the connubial act should be linked with an enormously pleasurable sensation. If She had not implanted this sensation in men and women, the human species would surely have perished.'[42] De Graaf also disagreed with Aristotle when it came to the function of the male testicles, which the Greek philosopher argued were not necessary for generation, despite the fact that castrated animals were normally sterile. His explanation was simple, if odd: the testicles were supposed to act as weights that helped the movement of semen. De Graaf dismissed this idea, and argued on the basis of his dissections that the male testicles produce semen. However, De Graaf accepted the thoroughly Aristotelian idea that semen is produced from blood. But having failed to find anatomical evidence to support the various theories as to how this might take place, he had to come up with his own vague idea, which involved the action of 'animal spirits' flowing down nerves in the testicles and fusing with the man's blood.

De Graaf's debt to Aristotle was at its clearest when he dealt briefly with the question of generation. Starting with Harvey's declaration that no semen can be found in the uterus after copulation, De Graaf came up with the following male semen-centred vision of generation: 'The virtue or more subtle part of the semen causes a fermentation in the membranes of the uterus. As a result of this fermentation there afterwards exudes from the membranes of the uterus a fluid, which, wrapped in a membrane, contains the rudiment of the foetus.'[43] This was simply Aristotle's view of the male's semen as the decisive factor in generation, dressed up in some of Sylvius's ideas about 'fermentation'.

So while De Graaf's dissections of the male organs of generation were innovatively accurate and precise, and his attitude towards women and their sexuality was apparently quite modern, his understanding of the role of the female in generation was entirely traditional: it was the male's semen alone that was 'a body full of spirit capable of generating a soul'.[44] The possibility that the female played an equal part in generation, even in terms of Galen's 'two-semen' theory, was not part of his world view. At this stage he saw the male's semen as the key factor – he did not even refer to Van Horne's and Steno's claim that women have eggs. At the end of his book, De Graaf said: 'If we learn that our labours have not been unwelcome, we shall later, provided time and opportunity allow, write another treatise about the genital parts of women.'[45] However, everything he said about generation in the pages of *De Vivorum Organis Generationi* shows that he had not yet begun that work.

De Graaf's book not only provoked a very public conflict with Van Horne, it also led to an unexpected and nasty private polemic with an eminent English scientist. An important part of De Graaf's study dealt with the structure of the male testicle, which had previously been thought to be a solid gland. De Graaf showed that it was in fact made up of an incredibly densely packed network of very fine tubules. Although he was unable to determine what went on in these vessels, he assumed that they contained, or made, some component of the male semen. While he was in France, De Graaf had used bull's testicles as his model for investigating the structure of the male reproductive organs. Although bull's bollocks (as De Graaf called them) were readily available and suitably large, their particular form was not ideal for demonstrating the testicle's tubular structure. De Graaf had therefore looked around for a more appropriate subject for separating the tubules, and had finally settled on an unusual alternative: the edible dormouse.

The edible dormouse, or 'gliris', has a bushy tail and looks vaguely like a small squirrel. It weighs around 100g, with its testicles weighing about 1g each. While the small size of the gliris testicle obviously posed technical problems that the bull's testicle did not, it was very easy to show its tubular structure. De Graaf removed the exterior membrane of the testicle, then put it into water. With gentle agitation it simply came apart – 'it can be seen with absolute clarity that the substance of the testicles consists wholly of tubules,' wrote De Graaf in his book (in his *prodromus* Van Horne had also said that the testicle was nothing more than 'a collection of very tiny strings'). Even on this relatively minor point of his study, De Graaf revealed his concern about protecting his findings: 'I often demonstrated this to the physicians and surgeons of this city, some of whom gossiped so much about it that I became afraid lest, because of the laziness of the engraver, another should snatch from me the glory of this magnificent discovery before my treatise appeared with all the necessary figures.'[46] However, it was De Graaf who would soon be accused of stealing the credit that should have belonged to others, and even worse, of impugning the integrity of the Royal Society in London.

In June 1668, the *Philosophical Transactions* carried a long letter from one of the Society's members, Timothy Clarke.[47] Clarke, who had recently been appointed second physician-in-ordinary to the King, was a well-spoken man in his forties and a friend of Pepys ('a very pretty man, and very knowing,' said Pepys[48]) – indeed, it was Clarke who first suggested Pepys might join the Royal Society.[49] The bulk of Clarke's letter contested a report in the *Philosophical Transactions* which suggested that Jean-Baptiste Denis in Paris had been the first to carry out blood transfusions in animals and humans.[50] Clarke argued vigorously that a group of Oxford physicians, and in particular Richard Lower and Christopher Wren, deserved the credit instead: 'We in England received this discovery from no foreigner,' he spluttered.[51] At the end of his letter, Clarke changed his target and claimed that, together with Lower, he had dissected the 'seminal vesicles' before De Graaf, and had come to similar conclusions about their interconnections. In a key passage he then disputed De Graaf and Van Horne's discovery about the tubular structure of the testicle, claiming that this was 'known to us long ago'.[52]

Clarke probably had mixed motives in attacking De Graaf and Van Horne. There is no reason to doubt his claim that similar ideas were already circulating in England, but, given that nothing had been published, the force of his subsequent criticisms suggests something else was at work. The main target of his irritation was Denis and his claim to have been the first to carry out a blood transfusion. De Graaf and Denis had been close friends in Paris and had jointly carried out experiments (although not, it appears, on transfusion). Clarke may have been aware of this link and lumped the two men together. Another possibility, given the distinctly chauvinist tone of some of his arguments, is that he may have been badly affected by England's failure in the second Anglo-Dutch war, which had ended less than a year before. The war, which had been deliberately started by the English in 1665, saw an initial wave of Dutch defeats, including the English capture of the poorly defended North American colony of New Amsterdam, which was rapidly rebaptised 'New York'. But in June 1667, a group of Dutch ships sailed audaciously up the River Medway into Chatham, where they set fire to a number of vessels and made off with the English flagship, which they towed back to the Netherlands.[53] A few weeks later, the English settled for a hasty peace, retaining control of New Amsterdam, but having clearly lost a war they thought they could win. As an ex-Navy man, Clarke must have felt stung by this demonstration of Dutch supremacy.

In July 1668, the Secretary of the Royal Society, Henry Oldenburg, wrote to De Graaf, apparently asking for a response to Clarke's claims. De Graaf's reply, which was sent in October, was pretty pugnacious.[54] He said he was 'astonished' that the Royal Society had attributed his discovery to others, and that while some previous reports could be 'twisted like a waxen nose to say something about the substance of the testes', his study, which involved the presentation of detailed drawings, 'is not to be set at naught by your people'. De Graaf stated that 'it plainly appears that your countrymen not only did not understand the true substance of the testes before myself, but have not yet understood it correctly from my sketch', before implicitly criticising the chauvinist tone of Clarke's article: 'I feel sure your countrymen will allow to other nations as much renown for their discoveries as they in turn desire to receive from others for their own.'

Intrigued by the suggestion that the testicle was made up of fine tubes, the Royal Society had already instructed three of its members to try and 'unravel' a testis.[55] The initial experiments on rabbit and dog testicles were unsuccessful, with no long 'threads' being found, only short strands which turned into 'a mucous or slimy substance' when rubbed between the fingers.[56] By November 1668 a leading Fellow of the Royal Society, Dr King, had managed to unravel a human testicle, showing it to be composed of vessels, and had stuck the specimen 'on a glass' so that everyone could see.[57] Around the same time Clarke, prompted by Oldenburg, replied to De Graaf, attacking his letter as being 'filled with that petty ambition and youthful heat which renders young men obnoxious'. He then went further than before and claimed that the idea that the testicle is made of fine tubes 'has had diligent supporters among the English for more than twenty-five years and is at last almost well-known', while eye-witnesses could testify that eighteen years earlier in Padua 'a certain German' had demonstrated this. Confusingly, having seemed to accept that De Graaf was right about the structure of the testicle (which had been confirmed by King's recent experiment), Clarke then challenged De Graaf to prove it.

De Graaf's response, written to Oldenburg in February 1669, began in a restrained fashion, declaring that his letter had been mis-interpreted. He pointed out, quite legitimately, that even if previous studies had produced similar results, he could hardly be accused of having stolen those findings because nothing had been published. However, as he closed his letter, De Graaf showed his irritation, wondering 'whether it befits the discretion of so great a man to go on to say that ambition and youthful ardour shone forth in my letter, that I with little candour aim at a subterfuge in equivocal words, that I behave like a trickster, etc. I could have replied to all this in a similar style did I not venerate the age of that famous man, and did I not prefer to demonstrate from the facts themselves that I am not such a man as his prejudiced imagination feigns me to be.'[58]

This only annoyed Clarke even more, and in May he continued his dual attack on the young man, simultaneously claiming that De Graaf was not the first to describe the tubular structure of the testicles and contesting whether this was even true, sarcastically suggesting that one explanation might be that Dutchmen have differently shaped testicles to Englishmen. Oldenburg sent a copy of Clarke's criticism to De Graaf, accompanied by an unusually patronising letter in which he supported Clarke's suggestion that 'Mr De Graaf had accused of plagiary and injustice not individual Fellows of the Royal Society, but the whole Society', and pointedly reminded De Graaf that Van Horne also claimed to have been the first to observe the tubular structure of the testicle.[59] The serious nature of the allegations that were coming from London suggests that De Graaf had genuinely shocked the Fellows of the Royal Society and had not simply irritated them by committing the twin sins of being both right and young. However, not everyone agreed with Oldenburg's endorsement of Clarke's arguments. Samuel Colepresse, an English student at Leiden, who knew De Graaf, complained to Oldenburg that Clarke was 'a little too sharpe with De Graaf wch. strangers resent'.[60]

Apart from giving De Graaf a series of verbal cuffs round the ear, Oldenburg also provided the Dutchman with a way of resolving the dispute, by suggesting he send a dissected testicle to London. De Graaf seized the opportunity and apologised (a bit) in case he had been misunderstood; he then pursued the argument over the anatomical detail in more measured tones and sent to London 'the testicle of a dormouse, unravelled by my method'.[61] A drawing of this specimen − 'A little glass sent to Mr OLDENBURG from Dr DE GRAAF, containing a testiculus gliris unravelled and swimming in spirit of wine, designed to prove, that that organ is made up of nothing but small vessels'[62] − appeared in the *Philosophical Transactions* in October 1669, together with a report of King's experiments

*De Graaf's illustration of his dissection of a dormouse testicle, as published in
the* Philosophical Transactions.

which supported De Graaf's findings. A summary statement which
accompanied the drawing and the report suggested that the findings
proved 'that the Testes of Animals are made up of nothing but Vessels
and their liquors'.[63] After eighteen months, the question was more
or less settled. In December, Clarke accepted that De Graaf had not
plagiarised his findings and that he was right about the structure of
the testicle, nonetheless grumbling, 'I would have preferred to see
the testis of some larger animal, that is, of man, horse, bull, or boar
prepared in this way.'[64]

All these centuries later, the most striking point about the argument between Clarke and De Graaf is not the anatomical detail, nor
even the question of priority. What lingers in the mind is the furious tone emanating from both sides. Clarke was pompous and
condescending and used terms he later regretted,[65] but De Graaf, as

in his previous clash with Van Horne, was more than vigorous in his own defence. His hot temper rapidly burst to the surface and the debate turned into a polemic, although the public reputation of each side was saved by the fact that the row primarily took place in private correspondence which was published only in the 1960s. De Graaf had shown himself to be a skilled anatomist and, on two linked occasions – faced with both his one-time teacher and an eminent representative of the Royal Society – a determined advocate of his claims. He had also revealed his fears that 'spiteful fellows' might 'snatch from me the glory', suspicions that would have been reinforced by Clarke's deeply unfair allegations and which would resurface a few years later in rounds of claim and counterclaim of plagiarism and stolen priority, which saw him pitched against his one-time friend, Jan Swammerdam.

In the meantime, the growing focus on generation shifted away from the components of conception, to centre on the most complex and mysterious aspect of the whole process – embryonic growth and the way an organism develops from a seemingly featureless egg.

THE RULES AND THEOREMS
OF GENERATION

For seventeenth-century thinkers 'generation' was something that described both the means by which organisms appeared (spontaneous generation, egg-laying or live birth), and the way they grew after conception (development). Redi had shown that the most straightforward part of this set of questions, dealing with spontaneous generation, could be successfully studied using the tools and concepts available at the time. The complex issue of growth was another matter, but throughout the century ambitious or foolhardy physicians, philosophers and natural historians had tried to understand how an egg turns into an organism full of lumps and limbs.

Most of these investigations had focused on the development of the chicken egg, and had led to a general agreement on the sequence of events leading from conception to hatching. But there was no common understanding of how these changes took place, and in particular there was disagreement about where the chick came from in the first place – did generation involve the creation of something new, or merely the assembling or unpacking of existing structures that were somehow present in the egg? To make a decisive advance something more was needed than a good experiment, something that told observers about the nature of science itself. Understanding requires both experimental results and a theory that can explain those results: it is not enough to observe, you have to be able to make sense of what you have observed. In autumn 1669, about a year after Redi's book on the generation of insects had appeared, Jan Swammerdam provided such a general theory of generation, as a direct consequence of Thévenot's encouragement and advice. Swammerdam's ambition was to explain what he called 'the rules and theorems of generation' – the laws underlying the development of physical form. To do this, the young Dutchman had to make a stand against thousands of years of thinking, and dare to criticise some of the greatest minds of his century.

As in so many things, the Ancient Greeks were the first to try and understand embryonic development. In a fifth-century BC text subsequently attributed to Hippocrates, it was suggested that 'Everything in the embryo is formed simultaneously. All the limbs separate themselves at the same time and so grow, none comes before or after other, but those which are naturally bigger appear before the smaller, without being formed earlier.'[1] This bold description is so wrong that it suggests the author had never actually looked at a developing egg: far from everything appearing at once, the structures of the embryo in fact appear slowly, in a strict order.

Aristotle, writing a century later, showed that it was quite possible to observe what was really going on using nothing more than the naked eye and a sharp mind. Aristotle dissected chicken eggs at varying stages, and accurately divided chick development into three phases: after three days of incubation, the chick heart can be seen; by ten days old, all the parts of the body can be detected, while at twenty days the chick is covered in downy feathers and can cheep. Aristotle's largely correct descriptions of development are remarkable, given that he had neither microscope nor magnifying glass. However, the most important point was how he interpreted his findings – the principles he thought were involved in the growth and development of the embryo. He argued that there was an order underlying development which was due to different organs appearing at different moments: 'some of the parts are clearly to be seen in the embryo while others are not. And our failure to see them is not because they are too small; this is certain, because although the lung is larger in size than the heart it makes its appearance later in the original process of formation.'[2]

Although Aristotle had observed the slow, ordered pace of development, when it came to explaining where the parts came from and why they appeared in this order, he could only speculate, putting forward two apparently contradictory explanations. Firstly, he argued that the developing embryo might be like 'the automatic machines shown as curiosities. For the parts of such machines while at rest have a sort of potentiality of motion in them, and when any external force puts the first of them into motion, immediately the next is moved in actuality.'[3] For Aristotle, the semen provided the necessary 'motion' that set in train the sequence of events in the menstrual blood, leading to the appearance of the embryo.

But having used the compelling analogy of a machine, he went on to frame the whole process in terms of a non-mechanical explanation, which he called the 'final cause' – the 'end' or the 'purpose' of a given event. This idea is hard to understand today, as it involves a sort of circular reasoning that the scientific revolution drove out of our ways of thinking. For Aristotle the 'final cause' of embryonic development was to produce the adult organism. This meant that the ultimate explanation – the 'final cause' – as to why the various organs appeared in a given order was that this was the way to create a chick. This kind of argument has since become known as a 'teleological' explanation (from *telos*, meaning 'end'), and has proved a tempting dead end for many ideas about the living world over the past 2500 years. It appears to be an explanation, but on closer inspection you are no closer to understanding the processes and mechanisms involved.

By the middle of the seventeenth century, thinkers in a wide range of fields were increasingly abandoning Aristotle's 'final cause' explanations and were attracted to more direct, mechanical accounts of natural phenomena, including growth. The first of many attempts to understand the nature of development, using modern materialist explanations that went beyond Aristotle's views, was made by Descartes in the 1630s in his unpublished fragment *De la formation de l'animal*. But Descartes' overambitious mathematical vision of embryonic development was stillborn – even he was unsatisfied with it. At the end of the century, John Ray summed up the feelings of his contemporaries when he dismissed such mechanical attempts to explain generation as 'so excessively absurd and ridiculous, that they need no other Confutation than ha ha he'.[4]

In the middle of the century, with the country torn by civil war, three English thinkers produced a flurry of studies of generation and revolution. All three were prominent Royalists – Kenelm Digby (1603–65), Nathaniel Highmore (1613–85) and William Harvey. Each of them tried to surpass Aristotle's thinking about generation, and each of them, in different ways, failed. Digby and Highmore were both friends with Harvey (though not with each other), and all three men knew of each other's work – Digby referred to Harvey's deer dissections seven years before they were published, while Highmore had been a member of Harvey's Oxford scientific circle in the early years of the Civil War.

The first of the trio to publish was Digby – soldier, sailor, ambassador and pirate as well as alchemist, philosopher (he was a friend of Descartes) and son of one of the Gunpowder Plot conspirators.[5] In a theoretical book written in prison at the beginning of the Civil War and published in Paris in 1644, Digby claimed to have turned his back on explanations that involved 'occult and specifike qualities', preferring to 'shew that such effects may be performed by corporeall agents'.[6] Despite wanting to find a materialist explanation of generation, he disagreed with the idea that growth happens by assembling parts 'which were hidden in those bodies from when they extracted in generation'. That is, he opposed the suggestion that 'atoms' circulating in the body of the mother were assembled to form the embryo. Digby instead argued that generation occurs through 'the compounding of a seminary matter [semen], with the juice which accreweth to it from without'. For Digby, like Aristotle, semen transformed a pure 'homogeneall' substance – blood – into a living being. How this took place was not clear, but Digby insisted that before its transformation blood 'containeth not in it, any figure of the animal from which it is refined, or of the animal into which it hath a capacity to be turned'. Despite his initial intentions, he had done nothing more than slightly update Aristotle's vision of generation, according to which the semen shaped the menstrual blood into the embryo.

Digby's book led to a response from Highmore, who published *The History of Generation* in 1651.[7] Highmore was a convinced atomist, believing that all matter was made of particles, including those parts of matter that were responsible for generation. Unlike Digby, he had spent a great deal of time dissecting chicken embryos, and published his observations in a series of illustrations in his book. Highmore argued that blood contained particles responsible for generation which in some way represented the form of the animal. These particles were 'Atomes; which while they were parts of the blood, served for the nourishment and increase of that body from whence they were taken, but now serve to make up another Individuum of the same species'. According to this idea, later known as 'pangenesis', the particles responsible for generation were present throughout the body. Although wrong, this consistently materialist vision of generation set Highmore on what was basically the right track, when he argued that 'The conjunction of these seminal material Atomes of both Sexes, causeth this similitude of parts, and marks, with the parents that begot them.'

However, even this insight was not enough to enable Highmore to understand what he saw or to break completely with Aristotle. Highmore's vision of the importance of blood – reinforced by his dissections of chicken eggs, which showed that blood could be seen in the embryo before the heart appeared – bore the stamp of Aristotle's ideas. And when it came to the generation of insects and other animals such as eels and frogs, Highmore, like Digby, repeated

Aristotle's division of the natural world into 'perfect' and 'imperfect' animals, with the latter showing spontaneous generation. Despite the relative accuracy of his dissections of the egg, Highmore's explanation of the processes underlying what he saw was not very helpful: 'these seminal Atomes as soon as they are conjoyned in a convenient place, by the due ordering and regulating of the specifick soul, put themselves in order, fall to their proper places, and make up a Chick'. Just like that. What this 'specifick soul' was, and how it knew in what order to put the 'seminal Atomes', Highmore did not say. With its lack of precision and its invocation of an unspecified, unknown higher force, this had more than a whiff of Aristotle's 'final cause' about it.

A few weeks prior to the publication of Highmore's book, the most influential of the trio of English works on generation appeared. Following George Ent's pressure and flattery, the brilliant but flawed *Exercitationes de Generatione Animalium* emerged from Harvey's cupboard. As well as discussing how conception took place, much of Harvey's book dealt with the next stage – the 'development' aspect of generation. Like the material on conception, the chapters that focused on development were marked by impressive insights set in a confused theoretical framework. Harvey's study followed that of his teacher, the Italian surgeon Hieronymus Fabricius (1537–1619), which had been published in 1621.[8] By dissecting chicken embryos each day after conception, Harvey was able to correct his master and produce a precise description of the steps in the generation of the chick. Like Highmore, Harvey found that the first sign of life in the fertilised egg was the appearance of blood, which could be seen twenty-four hours after conception. For Harvey, this was explained by the Aristotelian view that 'blood is the genital particle, the fountain of life, the first to live and the last to die, the chief habitation of the soul'.[9] He also investigated the role of heat in incubation, drawing attention to the Egyptian technique for incubating large numbers of eggs in heat-generating dung heaps.

However, the key point in Harvey's work on chick development was not his timing of the various stages, nor his interest in incubation, nor even his mystical, Aristotelian view of the role of blood. The most influential feature of this section was his attempt to provide a theoretical explanation of what happens during development. Unlike previous investigators of generation, Harvey preferred to try and explain what he could see – the gradual appearance of parts of the embryo – rather than speculate about what he could not. Following Aristotle, he noted that development involved the ordered appearance of the organs, apparently from nowhere. To describe this process he coined a new word, 'epigenesis' – 'the addition of parts budding out of one another'.

Harvey tried to understand the order that could be seen without bringing in a 'final cause'. Because he could not hope fully to comprehend what was going on, his explanation of how the various body parts appeared during this 'epigenesis' was both tantalisingly modern and infuriatingly vague. Like Digby and Aristotle, he was opposed to the atomist view that the organs were in some way present in the egg from the very beginning. However, Harvey recognised that the egg must contain some kind of representation of the future organism, and suggested that it was something in which 'there is no part of the future offspring actually in being, but all parts are indeed present in it potentially'.

Although this could seem quite correct – we would say the fertilised egg contains genetic information which enables the organs to form – Harvey was in fact trying to dodge a very difficult problem simply by stating the obvious. The early egg contains no detectable structures, and yet organs develop within it – they must come from somewhere. He rejected the atomist idea that there were tiny versions of the organs stored up in some part of the body, so the only possible explanation was that they must exist 'potentially'. Not surprisingly, Harvey gave no indication of how these potentially existing parts could be detected or studied, for the simple reason that observing a 'potential' seemed impossible.

Since science progresses by disproving existing hypotheses, for a new theory to become necessary old ideas must be unable to explain a particular event. If science lacks the means to test a particular hypothesis, due either to a lack of technology or a lack of alternative explanations that encourage experimentation, or both, that subject will stagnate. This is what seemed likely to happen with the 'development' aspect of generation: Harvey had come up with an explanation that, as it turned out, was partly correct. But his idea was at the same time so vague and so far-reaching that it threatened to stifle all future research. Unless someone could claim to see either Harvey's 'potentials' or the tiny particles or structures hypothesised by atomists such as Highmore, there was no way of testing Harvey's hypothesis.

Like all other thinkers before Redi, Harvey accepted Aristotle's division of the animal kingdom into perfect and imperfect animals, with the imperfect ones reproducing by spontaneous generation. Harvey added an extra layer to this idea: 'imperfect' animals also developed in a different way from perfect animals. 'Epigenesis', he claimed, could be observed only in 'the more perfect animals', while imperfect animals such as insects developed by what he called 'metamorphosis'. Nowadays this term is used to describe the still poorly understood process whereby insects with a larval stage (like butterflies, bees or beetles) are gradually transformed into the adult form via a stage taking place within a pupa or cocoon. Harvey's 'metamorphosis' had nothing in common with our meaning of this word. He used the term to mean an instantaneous and total change – the opposite of the gradual appearance of order or 'epigenesis' detected in the chick embryo. Harvey wrote, 'In generation by metamorphosis creatures are fashioned as it were by the imprint of a seal, or cast in a mould, that is the whole of the material being transformed.' He argued that in their transition from egg to adult, insects do not show slow growth, but rather the sudden and simultaneous appearance of all the bodily structures in their final form, 'fashioned as it were by the imprint of a seal'.

This idea was doubly attractive. Firstly, it developed the orthodox vision of how generation occurred without challenging it, thereby reinforcing Aristotle's division of the animal kingdom. Second, and most importantly for Harvey, it apparently corresponded to what he saw. When a caterpillar turns into a pupa it spins a silk cocoon, which soon hardens. Within an hour the hard, new, shiny exterior of the cocoon shows many of the key forms of the future adult – wings, antennae, legs, body – in a kind of relief. The same is basically true for beetles, wasps and bees, all of which Harvey studied. Both the tiny ridges on the surface of the pupa that outline the shape of the future insect, and the speed with which the transformation takes place, are in complete contrast to the slow appearance of barely discernible structures inside the chicken egg. Harvey was quite right: it looks very much as though the whole pupa has been stamped with a seal.

Furthermore, like other thinkers from Aristotle to Redi, Harvey argued that the pupa was a kind of egg. Many insects, it seemed, had two eggs – an 'imperfect' one which produced the larva or caterpillar, and a second 'perfect egg' which produced the adult. According to this view, both kinds of insect – the larva and the adult – were produced by generation from decay. The larva was generated by decaying matter, while the adult was generated by the decay of the larva or caterpillar within the pupa – 'out of which by means of a mere transformation all the parts are simultaneously constituted and embodied, and at last a perfect animal is born,' said Harvey. Larva and adult were seen not as two distinct stages of the same life form but as two separate organisms, each arising spontaneously from decay.

Harvey's contemporaries considered that his most important contribution to the study of generation was his work on chicks and on deer. In many ways, this was true – these were fields in which he made bold experimental attempts to resolve fundamental problems related to conception and development. But in terms of the influence of his work   its encouragement of further research, its effect on the progress of science – his development of the traditional Aristotelian view of insect generation proved to be the most important, and the most mistaken.

Overall, Harvey, Digby and Highmore had shown the importance of the developmental aspect of generation, but none of them had been able to make a decisive advance, partly because the processes they were studying are amazingly complex – they are still far from fully understood more than 350 years later – and partly because the important insights that could be gleaned with the techniques and theories that were available at the time would not come from chickens.

Many thinkers had been intrigued by the various patterns of growth shown by insects, some of which hatch from an egg and then grow slowly, others of which start as a caterpillar or maggot, turn into a pupa and then emerge as a creature with a completely different shape and behaviour. Understanding how these steps took place, and even appreciating what now appears to be the straightforward relation between maggot, pupa and adult, was a remarkably complex challenge.

At the end of the sixteenth century a series of attempts had been made to understand development in insects. The English physician, naturalist and Member of Parliament Thomas Mouffet (otherwise Moffett or Mouset or Muffet or Muffett – spelling had yet to be frozen by printing[10]) gathered together material by various authors which was eventually published in 1634, thirty years after his death, under the odd title *Insectorum sive Minimorum Animalium Theatrum* ('The theatre of insects or lesser animals').[11] The *Theatrum* is a collection of descriptions of the life and behaviour of a wide variety of insects – bees, wasps, flies, ants, butterflies, locusts – with a strong emphasis on their usefulness in medicine. It contains hundreds of stunning woodcuts of insects that pepper the text and wriggle all over the title page, many of which were apparently drawn – perhaps by Mouffet – with the aid of 'lenticular optick Glasses of crystal' (magnifying glasses). The book is marked by an uncritical acceptance of the views of ancient writers, even when their opinions were completely contradictory. For example, Mouffet apparently agreed with both Aristotle, who thought that bees were generated from dead animals, and with Pliny, who claimed that bees shed their 'seed' on to honeysuckle, then brought it back to the hive, where 'by diligent and soft sitting upon it, it comes to perfection'.[12] Part of the reason for this equivocation was that Mouffet was keener on book learning than he was on practical study, and much of the book was not in fact his own work, but had been taken, with acknowledgement, from a number of other sixteenth-century writers.[13] The end result was useful as a compilation of contemporary ideas, but not as an example of clear scientific thinking.

This was not Mouffet's only work on insect generation: in 1599, in order to encourage the growth of the silk industry in England, he had published a long poem on the art and science of silkworm husbandry, *The Silkewormes and Their Flies*.[14] Mouffet's poem dealt mainly with the mythology surrounding silkworms, together with practical questions about feeding them and so on, but it also provided some insight into contemporary views about generation. Although he recognised that silkworm caterpillars come from eggs which are laid by adults following mating, Mouffet's view of generation as expressed in his poem combined an alchemical vision of the eternal cycling of matter, with the idea that 'heat' – either literally, in terms of incubation, or metaphorically, in terms of some bodily quality – was essential to development.

Later in the seventeenth century, the 'Thrice Noble, Illustrious, and most Excellent Princess, the Duchess of Newcastle' (Margaret Cavendish) also wrote a poem about the silkworm, as part of her collection, *Natures Pictures Drawn by Fancies Pencil To the Life*.[15] Although a brief explanation placed after the poem gave a reasonably accurate description of the silkworm lifecycle ('The Silk-worm is first a small Seed; then turneth into a Worm; at last grows to have Wings like a Flye'), the Duchess's work, which at twelve lines was somewhat shorter than Mouffet's effort, presented a confused and classical image of several butterflies arising out of the decay of a single dead caterpillar:

> The Silk-worm digs her Grave as she doth spin,
> And makes her Winding-sheet to lap her in:
> And from her Bowels takes a heap of Silk,
> Which on her Body as a Tomb is built:

> Out of her ashes do her young ones rise;
> Having bequeath'd her Life to them, she dyes.

A clearer view of insect pupation came from Andreas Libavius (d.1616), a German alchemist and physician, and Ulysses Aldrovandus (1522–1605), an Italian naturalist who compiled a series of encyclopaedic studies of animals. Both men studied the silkworm, but with a more scientific and less poetic approach than Mouffet or the Duchess of Newcastle. In 1599 Libavius published a brief study of the life of the silkworm in his book *Singularium Pars Secunda* ('Remarkable things – second part'), in which he daringly suggested that the caterpillar and the butterfly were the same organism.[16] However, he had no experimental proof for his hypothesis, and his proposition went unnoticed for nearly seventy years. A more traditional line was taken a few years later when Aldrovandus, by now an old man of eighty-two, published his encyclopaedia of insects, *De Animalibus Insectis Libri* ('Book of insect animals'). This huge volume included some sophisticated woodcut illustrations showing the caterpillar's silk gland, the pupa and the moth hatching out. However, Aldrovandus did not understand the relation between the caterpillar and the moth, and neither he nor Libavius tackled the complex question of the processes underlying the transformation of one stage into the other.

This level of ignorance about the generation of silkworms might seem surprising, given the long-lasting economic importance of this insect – it had been cultivated for more than three thousand years, first in China and the Far East, then in Europe from ancient times.[17] However, for silkworm farmers down the centuries and across the continents, the key thing was to know which plants the caterpillars liked and which they did not, not whether the moth and the caterpillar were the same organism. A summary of contemporary views of silkworm generation was given in 1622 by a Frenchman, John Bonoeil, in a pamphlet designed to encourage the colonists in Virginia and the 'Summer-Islands' (Bermuda) to stop cultivating tobacco and to farm silkworms instead, and written with the encouragement of anti-smoking campaigner King James I. Bonoeil's *Treatise of the Art of Making Silke*, a manual for would-be silkworm farmers, outlined the necessary techniques for cultivating caterpillars, harvesting silk, and for ensuring that there were sufficient caterpillars for the next season. But Bonoeil's account was striking for what it left out – the charmingly inaccurate drawings of mating moths that accompanied the text showed that caterpillars came from eggs and turned into butterflies, which in turn produced more eggs, but Bonoeil made no attempt to place his observations in any kind of theoretical framework, in terms of either reproduction or development. As it turned out, the Virginia mulberry trees were not appropriate for silkworms, while tobacco was far more profitable than silk. The Virginia silk industry never took off, and the rest is history.

In the Dutch Republic fascination with insect transformation extended beyond the frontiers of natural history, deep into cultural and artistic life, playing a minor but striking role in the history of art. At the beginning of the seventeenth century, the centre of this fusion of art and science was Middelburg, in the south-west of the Netherlands. Middelburg grew fat on the activities of the Dutch East India Company; although it had no university and no famous philosophers or scientists, its wealth and commercial influence helped provide the impetus for the discovery of the telescope and also encouraged the development of a specialised genre of still-life painting – Dutch flower painting. This combined a strikingly colourful composition of varied blooms with a very precise *trompe l'oeil* style in which the flowers were rendered with hyper-real intensity. These were photographs in an era before photographs, a style that later in the century was transformed and applied to scenes of everyday life, reaching its peak in the work of Vermeer. To reinforce the impression of reality, Middelburg artists would add meticulously painted insects to their works – caterpillars, butterflies, bees and flies.[18]

One of the many Middelburg artists producing still-life paintings in the middle decades of the seventeenth century was Johannes

Goedaert (c. 1617–68). Like other artists, Goedaert often included butterflies and moths in his flower paintings. But as well as representing the striking form and colour of these insects, he tried to understand their natural history. Throughout his life, Goedaert was obsessed with tracking the transformation from egg to caterpillar to pupa to adult. He carefully collected larvae and caterpillars from the Zeeland fields, brought them home, did his best to find appropriate food to keep them alive, drew the caterpillar, noted when the insect began to pupate, drew the pupa, then waited to see what would emerge. This sometimes took several weeks, after which he drew the adult that he found fluttering, buzzing or scrabbling around in the container.[19] In 1662, after forty-five years of observations, but with no scientific or academic training, Goedaert published *Metamorphosis et Historia Naturalis Insectorum* ('The change and natural history of insects').[20] This study, published in both Dutch and Latin, was aimed at several different audiences: as well as natural historians, Goedaert also intended to interest gardeners, embroiderers and potential buyers of his paintings.[21] In 236 pages and eighty hand-coloured plates, he showed larvae, pupae and adults. In 1667 and 1668, two more volumes followed, containing a further 126 coloured plates.

The stunning illustrations in Goedaert's book, backed up by a series of observations (he never described anything resembling an experiment which would have enabled him to choose between possible interpretations), showed there was a link between the two stages of the life cycle of many insects – the larva (or maggot or caterpillar) and the adult. However, although Goedaert realised there was a consistent relationship between the two stages – for example, each kind of caterpillar produced a specifically shaped pupa which went on to release a specific kind of butterfly – he did not think that they were the same organism with different shapes. Instead, he accepted the generally held view that the adult was generated from the decay of the caterpillar.

To his surprise, however, Goedaert frequently found that a given kind of caterpillar did not always produce the expected butterfly – sometimes a caterpillar would generate dozens of flies or wasps

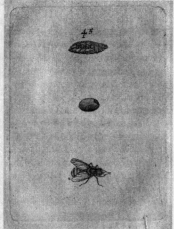

Two illustrations from Goedaert's Metamorphosis et Historia Naturalis Insectorum. *The caterpillar on the left apparently eats 'rotten* Willow, *and is found lying in the Bodies of those Trees' before turning into the moth. Goedaert noted that the maggot on the right 'Eats* Violetleaves, *it can hardly endure the Sun beames, and therefore gets under Ground, it can hardly be found because it is green, like the leaves it feeds on'. He recorded that 'It changed, the* 4th. *of September, and abode in that condition without Meat or motion, to the* 9th. *of May, the Year following, at which time came forth . . . an ordinary* Flesh-fly.' *He concluded that 'the* Aurelia [maggot] *was* Carrion *and putrid, when the Fly fed upon it'.*

('out of one and the same *Species* of Catterpillars, a Butterfly is produced, and 82 Flyes').[22] He was nonplussed by this apparent violation of the systematic transformation of each kind of caterpillar into a specific type of butterfly; he felt that it was 'against the usuall course of Nature, that from one and the same Species of Animals, an Offspring of different Species shou'd be gendred'. But Goedaert had not observed closely enough; had he done so he would have realised that these tiny flies or wasps came from eggs that had been laid in or on the caterpillar and, once hatched, the larvae had eaten their prey before going on to transform themselves into their own adult form.

As to where caterpillars themselves came from, Goedaert, writing before Redi, believed in spontaneous generation. For example, he thought caterpillars found in fruit tree blossoms were 'bread by a moist winde . . . the watter out of which they are bread, is a moist cloud like Honey dew'. He even did a similar experiment to Redi, putting dead caterpillars in a jar to see what they generated, but lacking the Italian's rigorous logic and experimental spirit, the elderly Dutchman simply did not understand what he had seen: 'And when I did stop up their Dead Bodys in Juggs,' he wrote, 'I had a multitude of little Flyes, which I say without doubt were Generated of heat and corrupted matter.' Despite these mistakes, which were typical of the time, Goedaert's work showed that even if insect generation might be spontaneous, there was nothing random about it – a given kind of caterpillar would always produce the same kind of butterfly or moth, unless of course it produced flies or small wasps.

By the middle of the 1660s, our understanding of the nature of insect growth and transformation had barely progressed beyond that of Aristotle. Then, at a meeting of the Thévenot *académie* in Paris, sometime in 1665, a remarkable event occurred which changed the way that development was seen and showed that the growth of insects and of larger animals could be understood in the same way. During a discussion of animal generation which focused on 'how the viscera are laid down', Thévenot invited each of his guests to give their opinion. After everyone else had spoken, Swammerdam pointed out that 'the matter should be attacked not by thought but by experimentation',[23] and proposed to show the assembly what he meant at the next meeting. A week later, he brought along a large silkworm – a bit smaller than your little finger – that was about to pupate, and asked everyone to look for the wings of the future butterfly. As the creamy caterpillar slowly moved around, everyone agreed that there were no wings to be seen. Swammerdam then took a scalpel and cut a slit starting just behind the head, which extended the length of the body. He peeled back the caterpillar's cuticle and to everyone's amazement revealed the wings, head and proboscis of the future adult, very soft, pale, delicate and not fully formed, but clearly present.

With this audacious act of dissection, a small piece of the generation jigsaw suddenly slotted into place. Swammerdam had shown that the adult butterfly did not grow from the decay of the caterpillar but instead appeared in rudimentary form long before pupation began, suggesting that the two stages were in fact one and the same organism. Furthermore, the fact that the adult organs detected in the caterpillar were incomplete, soft and 'watery' and appeared so long before the butterfly hatched out of the pupal case showed that development in insects did not take place rapidly and completely, as Harvey argued. Insect growth was based on something very different from Harvey's vision of 'metamorphosis', something which in many ways resembled the slow growth of vertebrates.

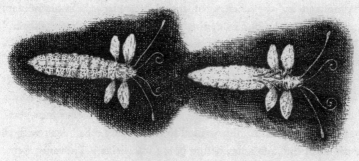

Swammerdam's drawings of a silkworm caterpillar from Historia Insectorum Generalis *(1669), dissected just before pupation, and showing adult structures (wings, antennae and legs). The two views show the caterpillar's back (left) and belly (right).*

The pieces of the future adult discovered by Swammerdam (they are now known as 'imaginal discs') are present in all insects that have a pupal stage; however, only in butterflies and moths do they look anything like the adult organs – in fly maggots, for example, they are just blobs of cells which require modern staining techniques, and a great deal of imagination, to see as future wings or legs. Had Swammerdam started off by dissecting a fly maggot, he would not have found evidence he could understand, and much of this story might have been very different.

Swammerdam's demonstration was so stunning that Thévenot was still describing it in letters four years later,[24] while the young Dutchman turned it into a kind of party trick which he performed for well-heeled visitors to his father's cabinet, including Magalotti, who had accompanied Prince Cosimo to Amsterdam during his tour of northern Europe in 1668–9.[25] Swammerdam was keen to show his radical discovery, and to counter the suggestion by Goedaert and Mouffet that the caterpillar and the adult grew in completely different orientations (as Goedaert put it, 'where the back of the Catterpillar was, there are the belly and feet of that Animal it's changed into; and the contrary, where the belly and feet of the Catterpillar were, there now the back of that Animal is, which was produced by the change of the Catterpillar').[26] Swammerdam showed that the bodies of the butterfly and of the caterpillar had exactly the same orientation, for the simple reason that they were the same organism.

Thévenot encouraged his protégé to study the generation of other insects, and to gather his material together and publish it. However, Swammerdam did not produce his account for another four years. The perfectionist in him wanted to make sure he had dealt with all the most difficult issues before going into print – Thévenot wrote that: 'before beginning the printing and the writing of the natural history of insects, he wanted to use his own senses to clarify the generation of Bees, which remains most obscure, despite the fact that it is the subject on which the Ancients worked the hardest'.[27] And he also had to complete his doctoral thesis on respiration. By the time the dissertation was submitted in February 1667, Swammerdam's attention was clearly focused on the question of generation, as shown by the mating snails on the frontispiece, and above all by a promise he made in the preface, which caught the attention of a reviewer of the published edition: 'this *Author* in his *Preface* promises the publishing of a *Treatise* about *Insects*, in which he ingages to shew many wonderfull things in those little and seemingly contemptible Creatures, and in particular to demonstrate to the Eye the very method and manner how a *Caterpiller* is transmuted into a *Chrysalis*.[28] The final reason for Swammerdam's delay was more serious: sometime in the summer of 1667 he fell sick with a 'tertian fever' – almost certainly malaria – which eventually killed him, in 1680. For the rest of his life he repeatedly suffered from severe bouts of the disease, which made him very ill and prevented him from working for weeks on end.

Because of all these factors, Swammerdam's book on generation – *Historia Insectorum Generalis* ('A general history of insects') – did not appear until the end of 1669, although rumours of its imminent publication were already circulating in the scientific community in September 1667.[29] The title was doubly deceptive: not only was the book written in Dutch rather than Latin, it was anything but a 'general history' of insects. Building on his silkworm dissection of four years earlier, Swammerdam presented an immensely innovative description of the generation of insects and other animals. But *Historia Insectorum Generalis* had none of the style and panache of Redi's book. Despite occasional lyrical passages singing the praises of the anatomy of insects and the order he could detect in his observations, Swammerdam was not a poet nor was he trained in courtly rhetoric. His book was repetitious and contained none of the strictly logical demonstrations that marked Redi's masterpiece. This was only partly because of a difference in the two men's literary talent; the nature of Swammerdam's subject meant that it was much more difficult to set out a strong case following clearly defined lines and backed up by rigorous experimental evidence.

Swammerdam's book may have been slightly disorganised, but it was extremely ambitious – he thought his study would ultimately reveal the laws of growth that were common to all animals: 'Nature performs the whole process of generation in these insects in so clear and open a manner, that by the assistance thereof it should seem as if we could penetrate into the true foundations, hitherto buried in darkness, of the generation of other animals, which we shall evidently demonstrate, when we have time and opportunity for further experiments.'[30] But understanding generation in insects was not straightforward. In most insect species, it was difficult to understand development because it was not clear exactly what a pupa or cocoon was. Some people thought it was some kind of casing for the larva or caterpillar to die in, which would then contain the resulting decay until the adult was spontaneously generated. Both Harvey and Redi were more sophisticated, arguing that the pupa was 'an egg', implying that insects simply went through the same process twice over. But Swammerdam's silkworm dissection had shown that the pupa was not an egg of any kind, but something much more surprising – it 'not only contains all the parts of the future animal, but is indeed that animal itself'.[31] To be more precise, it 'is nothing more than a change of the Caterpillar or worm . . . containing the embryo of the winged animal that is to proceed from it'.

Swammerdam's encyclopaedic knowledge of insects, acquired though his endless collecting, coupled with his appetite for finding common rules underlying generation, led him to realise that all 'insects' – this term included such non-insects as scorpions, spiders and even frogs – could be placed in four 'classes', each corresponding to a different mode of transformation or development. For Swammerdam, each stage involved what he called a 'nymph' – a phase of growth in which the adult form appeared – which could take place in the egg, as part of a free-living intermediate stage, or in a cocoon. The four groups he outlined were both a description of what he observed and an attempt to provide a theory of how growth took place.

The first type included lice, earthworms, leeches and snails, all of which hatched out of the egg fully formed – they grew with age but did not change their shape. The second class was made up of those insects that went through a series of moults, changing slightly each time, and included dragonflies, grasshoppers, locusts, crickets, cicadas, cockroaches, water striders, water scorpions, mayflies and earwigs. Swammerdam's third group, derived from his studies of the silkworm, consisted of insects with a pupal stage in which the form of the future insect could be seen, and included butterflies and moths as well as bees, beetles, wasps, the wolf fly, hornets, bumblebees, mosquitoes, various 'flies' that look like butterflies or scorpions, ants and stag beetles. The fourth class, which included flies, tested his ingenuity to the limit. Unlike butterflies and moths, it was impossible for him to detect adult structures in fly larvae, and from the outside the pupa showed virtually no trace of the adult growing within. Swammerdam admitted that the changes shown by these insects 'seem indeed wholly incomprehensible',[32] and over the next few years he made a special effort to comprehend fly development, without ever succeeding – hardly surprising given that we still do not understand all the steps involved in the transformation of a maggot into a fly. Swammerdam's four classes are still used today to group insects, which gives an indication of the accuracy of his description. In fact, his idea exerted a powerful effect on the development of classification as a whole, because it showed that characters apart from shape or food sources could be used to classify species.[33]

Swammerdam also resolved two minor mysteries, each of which threatened to undermine his consistent vision of insect generation: the scores of tiny wasps that both Goedaert and Redi had seen emerging from caterpillars, and the origin of gall insects. The first problem was dispatched relatively simply: having observed insects laying eggs on or in caterpillars, Swammerdam rightly concluded that the larvae that hatched from these eggs 'conceal themselves in the bodies of living Caterpillars, out of which they again eat their way'.[34] The second difficulty required a bit more effort, but was eventually resolved. In 1666, Swammerdam wrote a letter to 'a close friend' in which he solved the very problem which, at around the same time, Redi was finding so difficult.[35] Swammerdam noticed that some insects would put 'seed or eggs' deep into plant tissue (he later made a striking drawing of this). If he kept the affected part of the plant, several months later insects appeared that were of the same kind as the individual that laid the eggs. He then dissected a number of insects and found that 'On opening the bodies of these Flies, we

Swammerdam's drawing of a gall wasp laying an egg in a bud. This drawing was done sometime in the 1670s, but was not published until 1737. The original is tiny: the size of a postage stamp.

meet with eggs, which perfectly resemble those found in the same excrescences; from whence, as well as from many other observations, we may fairly conclude, that all the Worms found in vegetable substances, are originally deposited there by the parent insects in the form of eggs.'[36] What Redi found so perplexing, despite dissecting tens of thousands of galls, Swammerdam was able to explain by some simple observations.

After studying the classic cases of generation in so many different species, as well as explaining the two challenging counter-examples of parasitoids and gall insects, Swammerdam summarised his findings in an aphorism which echoed Redi's conclusion and which contained the key to a scientific understanding of generation in all organisms: 'all insects proceed from an egg, that is laid by an insect of the same species'.[37] Not only was this far more precise than the '*Ex ovo omnia*' found on the frontispiece of Harvey's book on generation – by 'egg' Swammerdam meant something not too far from what we mean today – it also showed where the egg came from and said that like bred like. With this statement, Swammerdam propelled the whole of natural history into the modern world, allowing future thinkers to investigate the nature of species – groups of organisms which breed true. Over the next 150 years, this view of species as fixed groups slowly gave way to various theories of evolution, in which species were seen to be separate from each other, but able to change gradually over time. Without proof that species breed true, the modern, Darwinian idea of evolution could never have taken hold.

Swammerdam was not content simply to describe generation; he wanted to understand how it occurred. To do so, he had to deal with the overwhelming influence of Harvey's book and the striking difference between his own findings and those of Harvey when it came to the generation of insects. After praising Harvey, calling him 'the second Democritus', Swammerdam quoted his admittedly poor description of the generation of bees and scornfully concluded: 'his dissertation contains almost as many errors as words. This is even more surprising in one so well versed in enquiries of this kind, where truth can only be ascertained by experiment!'[38] Although this is fairly muscular, not only was Swammerdam basically correct on this point, his robust style was typical for the time.

Swammerdam summed up his criticism of Harvey's view of insect growth as follows: 'all the limbs of the butterfly, the fly, and other such insects, do actually grow in the worm, and in the same manner as the limbs of other animals . . . Those parts are by no means generated suddenly and all at once, as has been supposed, but grow leisurely one after another under the skin that covers them.'[39] In other words, Swammerdam argued that all animals develop slowly, obeying the same rules of growth that Harvey had described in the chick egg. Setting out those rules was not easy, however, as shown by Swammerdam's groping attempt to describe what happens during insect development: 'swelling, sprouting, protruding, ripening and budding, and the appearance of new limbs, which are the result of the fattening of parts, and in no way the result of reshaping'.[40] This is a striking description of what Harvey called 'epigenesis' – the slow growth of parts in a specific order – and was completely different from the near-instantaneous 'metamorphosis' he claimed occurred in insects. Far from opposing Harvey's use of epigenesis to describe development, Swammerdam extended it to insects. He then went even further and pointed to the gradual but complete change of a tadpole into a frog, and the growth of a plant from a seed, to show that there was nothing mysterious or unusual about the developmental changes shown by insects: all forms of life, from insects to amphibians to plants, grew in the same way. Rejecting Aristotle's pre-scientific division of the natural world into 'perfect' and 'imperfect' organisms, Swammerdam argued that all life obeyed the same rules: 'in respect to the accretion of their limbs there is not the least difference to be found between these large creatures, and the little worm we have compared with vegetable substances'.[41]

But even if the process of generation was essentially the same wherever you looked, the reasons for the way the adult parts of an insect grew in size, strength and complexity during pupation remained a mystery. Swammerdam came up with an explanation that was ingenious, which flowed from the evidence and corresponded to the most advanced thinking of the time − but was completely wrong. Having noted that the adult structures found before or during pupation were 'fluid, in a manner like water itself'[42] and that bee nymphs could 'weigh twice as much as the Bees that are produced from them',[43] he provided a straightforward mechanical account of what must be happening: the adult structures finally became hard and fully formed through a process of 'insensible perspiration',[44] or 'a slow evaporation of the superfluous moisture'.[45] Although Swammerdam could not know it, the change from pupa to adult in fact involves a complex pattern of decay, growth and differentiation. There is no evaporation involved.

Towards the end of his life, in the late 1670s, Swammerdam described a series of audacious studies on the frog which showed how generation takes place following fertilisation. Carefully written down on 104 gilt-edged pages, these manuscript observations, which were not published until 1737, dealt mainly with the frog's anatomy and physiology, but also gave a seventeenth-century view of the most astonishing phenomenon in the living world: the way a single

Swammerdam's attempts in the 1670s to clarify his ideas about the role of 'evaporation' in the formation of the adult insect inside the pupa, taken from the manuscript of his posthumous publication, The Book of Nature. *This part of the book was based on an edited version of his 1669 work,* Historia Insectorum Generalis, *as seen on the printed page on the left.*

fertilised cell turns into a complex individual. Swammerdam first tried to dissect frogspawn shortly after it had been laid; all that happened was that the small black dot was 'crushed, and otherwise disturbed by my handling it'. To get round this problem, he took some recently fertilised frogspawn and boiled it in water or soaked it in various 'liquors'. This produced hard, coloured eggs which he could study under his powerful single-lens microscopes, and even dissect.

What he saw was bewildering: 'I found it [the egg] to consist entirely of minute grains, which were in a manner uniformly divided, and were yellow and transparent, without any other contents or viscera, that I could discover. The little animal was also divided throughout, as it were, into two parts, by a very considerable furrow or fold.'[46] In other words, there was no small frog or tadpole in the egg, there were no discernible organs, nor any hint of order, apart from a single furrow dividing the structure in two, and count-less small 'grains'. Swammerdam's drawings show that there was a simple reason why he could see no organs: he had in fact observed the division of the newly fertilised frog embryo into two cells.[47] This is the first description of this fundamental step in the generation of every multicellular organism that has ever lived.

Swammerdam had a keen brain and a keener eye, and he was about to have a most perceptive idea. As he studied frog embryos of different ages, he noticed a 'sudden expansion and elongation of the young Frog's body, on the fourth day, when it explicates or unfolds itself'. This stage is marked by the development of key parts of the nervous system, and the appearance of the tadpole's tail and limb buds. Swammerdam now suggested that 'one part of the unfolded embryo forms the head and thorax of the future perfect animal, and the other part the abdomen and tail, which grows larger and larger by degrees'. This is exactly what happens, but it begs the question –

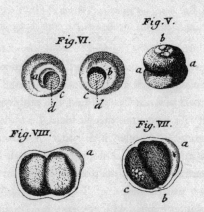

Swammerdam's drawings of frog eggs undergoing their first cell division. The egg in 'Fig. VI' has been cut in half along the line of the furrow – 'The broken substance of the young Frog's body, composed of grains dd, shewed itself in the place where these furrows terminated within,' reads the original caption.

which scientists are now beginning to answer – what decides which part is the head, and which part the tail?

In the pages of *Historia Insectorum Generalis*, Swammerdam gave an important summary of his views about development: 'in the whole nature of things there is no generation by accident, but by propagation and the growth of parts, in which chance has no part'.[48] This has been taken to mean that he believed that all the tiny parts of the future organism lay within the embryo, and simply enlarged to their final size. But Swammerdam never said anything so straightforward; his ideas about development were unclear and confused, despite his attempt to come up with a single theoretical explanation.[49] In fact, the key point he was making in this statement concerned his opposition to the role of 'accident' or 'chance'. Swammerdam was a mystical, pious man, and the avowed aim of his work was 'to make manifest, as clearly as I can, the stupendous works of the most adored and all-wise creator'.[50] Given that he thought 'The works of God are founded on constant and uniform rules',[51] he was deeply hostile to any intrusion of chance into the Universe, in the shape either of spontaneous generation or of apparently chance moments in development, because it would diminish the glory of God's creation. Indeed, Swammerdam's belief in the lawfulness of the Universe was extreme. In 1638, Sir Theodore Mayerne had written the Introduction to Mouffet's posthumous book on insects, and had used the generally accepted view that the change from a caterpillar to a butterfly was some kind of transformation as a justification for his alchemical views, asking rhetorically, 'And if Animals and Plants be transmuted, why should that be denied to Metals?' Swammerdam scornfully dismissed this suggestion, arguing that the existence of any kind of 'transmutation' would undermine the ordered glory of God's Universe.[52]

Swammerdam was a child of the seventeenth-century scientific revolution, not of the eighteenth-century Enlightenment. Discoveries of natural order reinforced his religious belief rather than undermined it, to the point that his scientific research and his religious attitudes became completely intertwined. At several points in his writings he suggested that accepting spontaneous generation opened the road to atheism, precisely because it implied there was no order in the Universe. He was therefore doubly hostile to this idea: it did not correspond to the facts, and it suggested a world in which God had not created perfection.

While *Historia Insectorum Generalis* was being printed, in mid-1669, Swammerdam received striking confirmation that his work on the silkworm was right. As he wrote in a paragraph hastily inserted into his manuscript: 'Whilst the preceding sheets were at the press, the incomparable anatomical observations of Dr. Marcellus Malpighius, professor of physic and philosophy, in Bologna, on the Silk-Worm, and its Butterfly, which the Royal Society of London, instituted to promote natural knowledge, caused to be published this year, 1669, were kindly sent to me by the noble Thévenot, whose merit and zeal to promote natural knowledge, are sufficiently known to all who happened to be at Paris, and present at the weekly disputations instituted by him.'[53] Swammerdam might have been pipped at the post, but he was generous in his praise for his Italian competitor Malpighi, pointing out that apart from Libavius he was 'the only person who excludes the fancied metamorphosis from the natural course of the changes, which the Silkworms undergo'. Swammerdam was not the only one to think that the link between the findings on insect generation in Malpighi's book *De Bombyce* and those in *Historia Insectorum Generalis* was of major importance. The unsigned 'account' of Swammerdam's book in the *Philosophical Transactions* underlined the fact that for Swammerdam 'the doctrine of Seigneur *Malpighi*, in his Dissertation de *Bombyce* (dedicated to the *R. Society*,) concerning the change of Butterflyes, is true'.[54]

Malpighi was a physician in his early forties who had recently taken up a post at Bologna and had been made a Fellow of the Royal Society following the publication of his book on the silkworm. He rose to fame in the early 1660s when he studied the frog lung and discovered capillaries – the fine blood vessels in the body's tissues. This provided the final proof of Harvey's hypothesis of the circulation of the blood; before Malpighi, scientists had not been able to show how the venous and arterial systems were connected, although they were certain that such a connection must exist.

Malpighi's interest in the silkworm was very different from Swammerdam's. The young Dutchman was fascinated by insects, but used the silkworm to prove something important about generation. Malpighi's interest was more general, and had been partly prompted not by his own curiosity, but by a letter he had received from London. At the end of 1667 Henry Oldenburg had invited Malpighi to write to the Royal Society about a number of subjects, including the silkworm. A little over a year later, Malpighi had sent Oldenburg the manuscript of his ground-breaking study of the anatomy and development of the silkworm – *Dissertatio Epistolica de Bombyce* ('Epistolary dissertation on the silkworm')[55] – in which he described the internal organs of an insect for the first time. When his book appeared in July 1669 – a little under sixty thousand words long, and accompanied by forty-eight stunningly detailed drawings – it surprised readers because it showed that this insect had internal organs that were as complex as those of any mammal or large vertebrate. Given that Aristotle's classification of insects as 'bloodless' and without internal structures was still widely accepted, Malpighi's work challenged a key part of thinking about the natural world.

It also showed how anatomy and experimentation could provide striking new discoveries. Malpighi carried out experiments to prove how the caterpillar respired, and unravelled the silk gland, showing how the precious thread was produced. He also dealt with generation, describing the way the moth formed out of the caterpillar. Swammerdam was overjoyed, because Malpighi's dissection of the silkworm and its pupa confirmed exactly what he had found: 'in the caterpillar, even before the cocoon is spun, the first vestiges of the wings are hidden beneath the second and third segments; in the head, the antenna can also be seen,' wrote Malpighi.[56]

Malpighi did not leave his study of silkworm generation there; in one daring experiment he actually tried to artificially inseminate silkworm eggs. He had noticed that semen is present in the 'uterus' of the female silkmoth after mating – in striking contrast to Harvey's assertion that semen could not be found in the uterus of deer after copulation (strangely, Malpighi did not appear to notice this vital difference, which was due to the fact that the female silkworm stores sperm after mating). To see whether the physical contact of semen and egg was necessary for fertilisation, Malpighi 'took some unfertilised eggs, wet them with semen taken either from the uterus or from the male's reservoirs', and then incubated them by the traditional method used on silkworm farms, which involved them being carried under the shirt of 'a young virgin'. Sadly, the eggs did not develop.[57]

With Swammerdam's and Malpighi's work, one part of the complex puzzle of generation had been resolved. It was now clear that all organisms obeyed identical laws in terms of their reproduction and growth: like bred like, even in gall insects and parasitoids, and all development was slow, involving the gradual unfolding of parts which, in some unknown way, were present from the moment of conception. Neither man went much further than this, but in the eighteenth century their limited insights were turned into a powerful but mistaken theory of development which would dominate biological thought for nearly two centuries. Before this view could triumph, however, the two components of sexual reproduction – egg and semen – had to be seen in a new light.

6

LIFE AND DEATH

Thursday, 9 October 1673. London. Autumn had set in early – the first sharp frosts had occurred earlier in the week, promising a cold winter. The dank odour of dying vegetation from the nearby fields mingled with the tang of sewage and the dusty smell of burnt-out buildings that still lingered, even seven years after the Great Fire. The narrow and filthy streets were clogged with carts carrying building materials and barrels of beer, street-sellers shouted their wares from the shelter of doorways, while smart folk sat in coffee houses and talked about books and money and war. The third Anglo–Dutch war had now been raging for eighteen months and recently the public had become convinced that the Anglo–French alliance against the Dutch was part of a plot to re-establish Catholicism and royal dictatorship in England. Between cups of coffee, there was much to discuss.

As the morning wore on, the mist drifted and turned into a fine rain. Just south of the building site at St Paul's, ten men were approaching the offices of William Brouncker, Commissioner for the Navy, for a meeting of the Council of the Royal Society.[1] They were wealthy, slightly puffy from rich food and fine claret, their coats made of good cloth. In his early fifties, Lord Brouncker was the President of the Royal Society and a man of earthly pleasures. He had an impressive double chin, droopy eyelids and slightly pouting lips, wore jewelled buckles on his shoes and dined with Nell Gwynn.[2] A few years earlier he had scandalised London society by openly living with his mistress, Mrs Abigail Williams, in his official apartments. One of Mrs Williams' most ferocious critics – at least in the private pages of his diary – had been Brouncker's friend and subordinate, Samuel Pepys, who was also making his way to the meeting.

There were no scientific issues on the Council's agenda that day, no thought-provoking experiments, no outlandish and unbelievable reports, just two technical questions. First there was a long discussion about where the Society would meet in the future – they were still looking for a permanent home following the destruction of their former headquarters in the Great Fire. Then the Council passed to the key issue of the day, a report from a small committee of three Fellows of the Royal Society, written on both sides of five sheets of paper in thin, neat handwriting. The archive copy of the report carries the following handwritten note:

> A letter addressed to the R. Society by three Physitians, their members, containing their Sense upon two bookes, dedicated to the Said Society by two Dutch physitians; the one, of Dr. Swammerdam, call'd *Miraculum mundi, sive Uteri muliebris Fabrica*; the other, of Dr. De Graaf, intitul'd *Partium Genitalium Defensio*; about which books the said Dutch Physitians had desired the R. Society's judgement especially as to the differences therein contained.[3]

At the heart of these 'differences' were the rival claims of the two former friends over the detailed anatomy of the female genitalia, and above all over who had been the first to discover the human egg. By setting up an informal committee to judge the issue, the Royal Society had wanted to give Swammerdam and De Graaf a way to end their quarrel, which had turned into a feud. The Society was looking to the future, to the discoveries the finding would inspire, and the potential contribution of these two brilliant young men to the development of knowledge. But Swammerdam and De Graaf had a different aim: before they could create the future, they wanted to settle their place in the past.

Steno and Van Horne had revolutionised our understanding of generation by suggesting that women have eggs. But although their idea was immediately and widely accepted, the new science required proof, not mere opinion or hypothesis. Neither man had reported any evidence, and it was unclear where it might come from. Van Horne was dead, Steno was apparently more interested in rocks and religion than in pursuing his studies of the human egg, while Swammerdam's approach to generation was clearly focused on insects.

Then, in February 1671 – just over a year after Van Horne's death – a 30-year-old Leiden graduate called Theodor Kerckring published a ten-page pamphlet, *Anthropogeniae Ichnographia, Sive Conformatio Foetus ab Ovo* – 'Drawings of the origin of man, or the constitution of the foetus from an egg'.[4] As its title implied, Kerckring's booklet contained a series of drawings, showing the female genital organs, embryos of various ages, and 'eggs' from humans and cows. Like Van Horne and Steno, Kerckring said the ovarian follicles were 'eggs', but he also insisted they could be found in both married women and virgins. He further claimed that two of his observations proved that the follicles were eggs: firstly, the 'eggs' clearly did not contain the kind of female 'semen' which people like Thomas Raynalde thought was generated in the female 'testicles': when the eggs were 'handled and slightly pressed, there will stick a little skin to the finger, which shews that 'tis not seed, nor anything like it,' wrote Kerckring. Second, the 'glutinous liquor' in the eggs would harden if it was heated, 'just as the White and Yolk in other Eggs,' he said. Like a true seventeenth-century scientist, Kerckring then ate the cooked 'liquor' to see whether it was indeed an egg. 'The taste of them is flat and unpleasant,' he reported, without making it clear whether he thought this supported or disproved his hypothesis.[5]

Kerckring's pamphlet initially received a lukewarm reception. The Parisian physician Abbé Jean Gallois pointed out that his key evidence was not conclusive – just because something was sticky and went hard on heating, that did not make it an egg, he wrote.[6] He went on to sum up many people's doubts: 'It remains to be seen if these vesicles, which are attached to the woman's body, become detached, and if the kind of egg within which the embryo develops is one of these detached vesicles. That is where the problem lies.'[7] Gallois was highlighting the same issue that Steno had raised back in 1667: how did the egg leave the ovary and make its way to the uterus? Gallois pointed out that no one had actually seen a 'detached' vesicle, either in humans or in an animal, and even if a vesicle did become 'detached', it was not clear how it could get into the uterus. Kerckring proposed that after fertilisation took place in the ovary, the egg descended into the uterus via what he called the 'ejaculatory vessels'. Unfortunately, these 'vessels' were in fact the ovarian ligaments, which simply hold the ovaries in place. They are solid, not hollow.

Most people, including Van Horne and Steno, rightly thought the eggs moved down the Fallopian tubes, but even here there was a major problem, as the 'tubes' are extremely narrow and certainly not wide enough for a vesicle the size of a small pea to pass through. Kerckring had pointed out that the follicles were flexible and could be flattened, but even so there was still not enough space in the tubes. Gallois suggested that the tubes might open at conception – he had observed large Fallopian tubes in a dead woman, as well as in a lioness – but even he was not convinced by this idea.

There was another reason why Kerckring encountered suspicion where Steno had been met with enthusiasm. Normally, illustrations added to the value and impact of a piece of scientific work. In Kerckring's case, the opposite, almost, was true. Most of the drawings in *Anthropogeniae Ichnographia* were simply unbelievable; for example, in a 3-day-old foetus – 'a little round mass of the bigness of a great black Cherry' – Kerckring showed what he claimed to have observed: a little man in which he could see 'the head as distinct

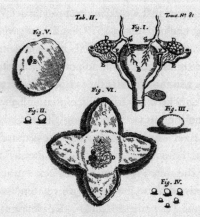

Kerckring's illustrations of eggs, foetuses and the uterus, as reproduced in the Philosophical Transactions. *'Fig. II' and 'Fig. IV' are 'eggs', 'Fig. V' shows a 3-day-old foetus, while 'Fig. VI' allegedly shows a '15-day-old' foetus.*

from the Body, and in the head we took notice of some traces of its principal organs', while for the 15-day-old foetus he drew a jelly baby with 'Eyes, Nose, Mouth and Ears, Arms and Feet'.[8]

This was fanciful nonsense – the 3-day-old embryo is composed of a mere eight cells and is the size of a pinhead, while the face takes four to eight weeks to develop. Even at the time, many of Kerckring's readers were doubtful. The reviewer in the *Philosophical Transactions* described his developmental claims as 'almost incredible';[9] Swammerdam, in restrained mood, merely said that his own studies showed that even at one month the foetus 'does not have the strength of the three-day-old foetus of Dr. Kerckring'.[10] Because they were so astonishing, Kerckring's illustrations were soon reproduced in the *Philosophical Transactions* and in the *Journal des Sçavans* – in a way, they were as shocking to those who agreed with the egg hypothesis as to those who were still opposed to it.[11]

But no matter how unlikely Kerckring's findings seemed, no one could confidently claim that they had more reliable alternative observations of human development at such an early stage.[12] A few years later, it was suggested that Kerckring had plagiarised Parisian surgeon Séverin Pineau's chillingly titled 1579 text *De Integratitis et Corruptionis Virginum Notis* ('On the signs of chastity and corruption in the virgin').[13] There is no evidence to support this accusation, beyond the fact that Pineau also thought that the very early embryo ('twelve days' in his case) looked like a jelly baby. It seems most likely that Kerckring was guilty of nothing more than having an overactive imagination; one thing is certain – he did not see what he claimed he saw.

Whatever its limitations, Kerckring's pamphlet underlined the importance of the new vision of human generation and the competition that existed to be the first to prove Steno and Van Horne's exciting and radical hypothesis. In Paris, Kerckring's publication was enthusiastically taken up by Jean-Baptiste Denis, who published a summary in French, along with his own comments, focusing on the new vision of generation created by recent research: 'All other animate creatures (not to speak now of Plants) are produced by the meanes of Eggs; as Birds, Insects of all sorts, Fishes (of which last sort though Whales, Sea-Calves and Dolphins bring forth live creatures of their kind, yet they first breed them within their Bodies in Eggs:) And why not Quadrupeds also and the Femals of Mankind?'[14] Denis subsequently went on to dissect cows in order to confirm Kerckring's finding and strengthen the case that mammals, including humans, came from eggs that were, or were found in, the small vesicles in the female 'testicle'.[15]

In London, the Royal Society reacted sluggishly. There was no attempt to repeat or test Kerckring's claim – indeed, there is no sign that it was ever even discussed. But as with the testicle affair, there was great interest in who had been the first to make such claims. An anonymous writer in the *Philosophical Transactions* summarised Kerckring's pamphlet shortly after it appeared, and hoped that 'that which hath been performed a good while since, upon this very curious and nice subject in *England* by the Learn'd Dr. *Timothy Clark*, one of his Majesties chief Physitians, will at last be made publick'.[16] However, whatever claim there might have been, it left no further trace in the historical record, and there was no repetition of the row over the male testicle – perhaps because Clarke died in February 1672.

Even before it appeared, Kerckring's work caused controversy. Throughout 1670, De Graaf had carried out a series of investigations into generation in rabbits, cows and women, linking anatomy and experimentation and focusing on the generation of humans from eggs. He later claimed that during this period he showed drawings of his work in progress to around two hundred people, including Swammerdam. At the beginning of 1671, Swammerdam heard about the imminent appearance of Kerckring's pamphlet and wrote to De Graaf warning him of the potential competition: 'I advise you to hasten the printing of your study of the organs of women, because I have reliably heard that Dr Kerckring is having printed something relating to this subject, and the figures are finished.'[17]

Given the explosion of rivalry, bitterness and spite which soon consumed the two men, Swammerdam's letter was strikingly friendly – he clearly did not feel he was in any kind of competition with De Graaf. At this time Swammerdam was still unsure what to do with his life: during the first half of 1670 he had been severely ill with another bout of malaria, followed by a long convalescence in Sloten, near Amsterdam, and his father had renewed pressure on him to start practising medicine. After much argument, Swammerdam had reluctantly decided to follow his father's wishes and abandon his studies of anatomy and entomology. Sometime after hearing about Kerckring's book, however, he evidently changed his mind.

De Graaf rashly dismissed Swammerdam's warning, saying that he had no intention of speeding up the publication of his book simply because of Kerckring. In fact, Swammerdam's piece of gossip was not the first time De Graaf had heard of potential rivals. At the beginning of 1670, Steno had visited him and the two men had discussed work on the egg in humans and other animals. During their conversation, Steno told his friend that he had found eggs in a range of mammals – fallow deer, guinea pigs, badgers, red deer, wolves, asses and mules. While De Graaf was pleased to hear this – it supported the growing consensus that all female animals had eggs – he may have been less happy when Steno told him that Malpighi was also carrying out a detailed anatomical study of the female reproductive organs. However, Steno's news was already a year old when De Graaf received it, and nothing had been heard from Malpighi, so De Graaf may not have felt under pressure.

But any nonchalance on De Graaf's part would have evaporated a few months later, in May 1671, when Swammerdam came to Delft on a social visit. As De Graaf later recalled, their conversation started badly, because Swammerdam 'made so many protestations of his friendship for me and promises to help me that I became suspicious and felt he wanted to cheat me'. If this recollection is accurate, it seems an excessive and unbalanced reaction – even more so given that a few months earlier Swammerdam had given some excellent and impartial advice. It seems more likely that De Graaf later reinterpreted Swammerdam's behaviour in the light of what he heard next: Swammerdam told him that Gerard Blaes, the man who had claimed credit for Steno's work a decade earlier, had asked him for a drawing of one of Van Horne's 1667 dissections. Blaes wanted to publish the illustrations of uterus and ovaries in a new edition of Thomas Bartholin's textbook on anatomy, which he was editing and which was to be printed by the Hack company in Leiden. After some pressure from Blaes, Swammerdam handed over his drawing – the same one that Van Horne had pestered him to complete during the second half of 1669. On hearing this, De Graaf became very upset, forcefully reminding Swammerdam that he was preparing an entire book on the subject. Swammerdam tried to explain that Blaes had nagged him incessantly, and that after all it was only a figure, but De Graaf was still furious.

To placate his friend, Swammerdam suggested that De Graaf instruct the printer to stop the presses so that at least the caption (or figure legend) containing the verbal description of the dissection would not be included in the book. The next day De Graaf rushed to Leiden, and asked Vandamme, the director of the Hack print shop, not to print the legend. Vandamme refused, saying that he needed the approval of Blaes. De Graaf, clearly shaken, decided to stake his claim by writing another *prodromus*, this time dealing with the female organs of generation. At the time, Hack had just finished printing a new edition of De Graaf's book on the pancreatic juices, which was ready for the binders. The eight pages of De Graaf's *prodromus*, addressed to Professor Schacht of Leiden, was incongruously inserted after the text, where the index was supposed to be.[18]

De Graaf was worried by Swammerdam's drawing not because he thought it could compete with the experimental evidence he had been gathering, but because of the central role played by descriptive anatomical discovery in science and medicine at the time. Although more than a hundred books of anatomy had been published in Europe since 1600, few of them were innovative; the vast majority repeated each other, copying previous mistakes and restating inadequate understanding.[19] As a result, there was no thorough and accurate account of the female genital organs. Precise anatomical descriptions were a key part of the scientific revolution's attempt to provide a material account of the Universe, and were the starting point for future therapeutic developments. This explains why what now appear as unimportant questions of detail, such as those that divided De Graaf and Van Horne, and De Graaf and Clarke, excited such passion in the seventeenth century. It also explains why De Graaf had to publish his *prodromus* as soon as possible – although Swammerdam's illustration might not be able to compete with his own experimental studies, it could upstage his anatomical account. Swammerdam's skill with the scalpel and the pencil, and his ability to preserve material, meant that his study, however brief, might well overshadow De Graaf's work.

De Graaf's *prodromus* focused on the anatomy of the female genital organs, including a brief description of the structure of the clitoris and a summary account of the 'eggs', which he described in similar terms to Van Horne and Kerckring, and which he argued descended into the uterus through the Fallopian tubes. In a decisive change from his previous view of generation, according to which semen fermented in the uterus to form the foetus out of the menstrual

208 R. DE GRAAF *De Succo Panc.*

no interdicere solemus, quartanariis quandoque concedendum esse. qualis autem diæta singulis febribus intermittentibus conveniat, ex iis quæ jam diximus quisque colligere poterit.

Proponeremus adhuc alia quædam remedia, tam febribus intermittentibus quam earum symptomatibus accommodata, nisi Clarissimus D. *Sylvius* in praxeos suæ parte prima jam edita eorum plurima ac præstantissima proposuisset.

FINIS.

INDEX

209

REGNERI DE GRAAF, MED. DELPHENSIS, EPISTOLA, AD *Virum Clarissimum* D. LUCAM SCHACHT, Medicinæ in Acad. Lugduno-Batava Professorem, DE PARTIBUS GENITALIBUS MULIERUM.

M*Iraris, vir Clarissime, quod Tractatum de Mulierum Organis Generationi inservientibus, cujus figuras quam accuratissimè delineatas jam per annum cum dimidio Tibi, aliisque curiosis quamplurimis videndas exhibui, necdum in lucem dederim; sed mirari proculdubio desines, postquam consideraveris occupationes, quibus Praxim Medicam exercentes à proposito sæpenumero divertuntur,*
* N *illis*

De Graaf's prodromus on the female genitalia, written as a letter to Professor Schacht of Leiden, and hastily inserted at the end of a 1671 edition of his book on the pancreatic juices.

blood, De Graaf now combined Sylvius's idea of 'fermentation' with Harvey's suggestion that the semen acted at a distance, like an odour, to provide a new view of generation: 'the eggs of the testicles become fertile when the seminal vapour rising up to the testicles from the uterus, through the Fallopian tubes, puts the eggs into a state of fermentation'.[20] De Graaf never acknowledged this important shift in his opinion.

On 15 June 1671, De Graaf sent Swammerdam a copy of his newly printed book, including the *prodromus*, and asked him to get Blaes to remove the contentious figure from future editions of the textbook. Swammerdam replied that, before he did anything of the sort, he wanted De Graaf to explain some of the language he had used in his *prodromus* – as in the exchange with Van Horne three years earlier, De Graaf had said that he was worried that 'those to whom I had shown my figures [might] gain credit from my work'. 'I would be annoyed if you included me amongst those people, for you also let me see your figures, out of friendship,' wrote Swammerdam, obviously hurt. He then introduced a sharper tone into the exchange, reminding De Graaf of the length of time that he, too, had been working on the problem of female generation: 'you should read the *prodromus* on the parts of generation written by Mr Van Horne, because our figures were made at that time. I say our figures, because we worked and dissected together.' Swammerdam may have felt this way from the outset, but, given his friendly letter to De Graaf a few months earlier, it seems more likely that he had been irritated by the tone of De Graaf's *prodromus* and by his friend's aggressive reaction to the Blaes affair, and that after the publication of Kerckring's pamphlet he realised that he, too, had a stake in the question of priority in the dissection of the female organs of generation.

De Graaf's later description of his reaction to this letter is telling: 'I made no reply, because his request seemed unjust, and because if he had been innocent he would not have believed I was speaking about him. After all, I had shown my figures to two hundred people, and if he felt guilty it was pointless trying to explain matters.' In other words, for De Graaf, Swammerdam's protests merely confirmed his guilt. From this point onwards, the chain of friendship that had bound the two men was broken. They had begun a conflict that would nearly destroy both of their reputations.

At the very end of 1671, Swammerdam published the illustration that had so worried De Graaf. However, it did not appear in Blaes's textbook, which was still in press, but as a single printed sheet, together with a brief figure legend which described how the original observations of both reproductive anatomy and human eggs had been made with Van Horne on 21 January 1667 – pre-dating the published claims of De Graaf by four years and of Steno by about six months. These discoveries were also explicitly linked with Van

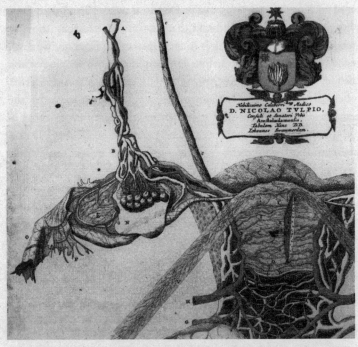

Detail of Swammerdam's drawing of his dissection of the human ovary and
uterus, from the printed sheet he sent to the Royal Society in 1671, dedicated
to Nicolas Tulp. The 'eggs' (in fact the follicles) are the grape-like structures.
The uterus is on the right, the feathery structure on the left is a Fallopian tube.

Horne's 1668 *prodromus* and, in typical seventeenth-century fashion,
the names of two reputable gentlemen – Drs Raei and Slade – were
invoked as witnesses, while the illustration was dedicated to one of
the most prestigious figures in Dutch medicine, Nicolas Tulp, who
thirty-five years earlier had been the dapper young surgeon at the
centre of Rembrandt's painting *The Anatomy Lesson of Dr Tulp*.

De Graaf had been right to be concerned. Swammerdam's illustration was, as its title proclaimed, an 'Exquisite demonstration' of
the structure of the female uterus. Beautifully drawn with delicate,
precise lines and accurate proportions, Swammerdam's sheet was

amazing: it was huge – about 40×30cm – and it was hand-coloured, with the arteries picked out in bright red.[21] Oddly enough, the most important part of the drawing was the least well executed. The portrayal of the ovaries was completely inaccurate: the left-hand ovary looked as though it contained a bunch of grapes hanging free; Swammerdam would in fact have seen very small follicles of varying sizes embedded in a solid, fleshy ovary. This was paradoxical in a publication that was designed to demonstrate his superiority in both dissection and representation. Like Kerckring's figures, it undermined the authority of its author, and De Graaf and others later seized upon this error. Swammerdam sent a copy of the figure to De Graaf, together with a bitter reproach: 'If you had replied to me as a friend instead of showing your disdain by silence, you would not today see this figure printed.' The letter went unanswered. Given that De Graaf's previous silence had led Swammerdam (rightly) to think he was being despised, this was not necessarily a good plan.

Before the row between the two men could grow any further, there was an additional indication of how widely accepted the egg hypothesis had become. At the beginning of 1672 a German physician, Christian Garmann, published a brief twenty-eight-page account entitled *Homo ex Ovo* ('Man comes from an egg').[22] Garmann discussed all the latest ideas about generation, including the writings of Kerckring and Steno, but set them in the context of a literary discussion about how the business of conception proceeded, in particular whether virgins produced eggs (he agreed with Kerckring that they could) and whether conception could take place outside the uterus, referring to Paracelsus's recipe for creating a 'chemical homunculus'. It is striking that the author of a text that was so steeped in pre-scientific ideas, quoting freely from Greek myths and the Bible, should have been so aware of, and so enthusiastic about, the latest scientific discoveries. Garmann, like many of his contemporaries, did not realise the contradictions that existed between the scientific material he quoted and the more literary interpretations of generation, nor the very different approaches to the natural world that underpinned the two kinds of account.

The next contribution to the investigation of human eggs could not have been more different. In March 1672, De Graaf published his masterpiece, *De Mulierum Organis Generationi Inservientibus Tractatus Novus* ('New treatise concerning the generative organs of women').[23] Three hundred and thirty-four pages long, with twenty-seven plates, the book focused on the form and functions of the female internal and external genitalia and, when it arrived in London a few months later, sold for five shillings.[24] Using dissections of humans, as well as of rabbits, hares, dogs, pigs, sheep and cows, De Graaf's work made a highly significant contribution to generation, by proving that something that was in the ovaries before mating turned into an embryo. That something was the egg.

As with De Graaf's work on the male genitalia, much of the anatomical detail is not of general interest. However, he did strike a blow against Galen's suggestion that the vagina is merely an inside-out penis – 'ridiculous', he called it.[25] This was part of a thread running through De Graaf's work – his attempt to explain the differences between the sexes, rather than defining the female as a deficient male, as had been done traditionally.[26] The material dealing with sex shows the ancient struggling with the modern. While De Graaf repeatedly suggested that the female external genitalia could be enlarged in those 'who, with lascivious thoughts, frisky fingers or instruments devised contrary to decent morals, wickedly stir themselves up',[27] he also recognised that the clitoris was the source of female sexual excitement. He thought it was of such importance that had it 'not been endowed with such an exquisite sensitivity to pleasure and passion, no woman would be willing to take upon herself the irksome nine-months-long business of gestation, the painful and often fatal process of expelling the foetus and the worrisome and care-ridden task of raising children'.[28]

The heart of De Graaf's revolutionary advance in the understanding of human generation was his study of the vesicles in the female 'testicles'. His drawings of these vesicles were far more realistic than Swammerdam's, and clearly showed how the follicles vary in size and position within the ovary. De Graaf did not claim to have discovered them – he presented a list of the anatomists who had previously noticed them, including such sixteenth-century giants as Vesalius and Fallopius (the discoverer of what we now call the Fallopian tubes). But none of these men had realised what the vesicles – or even the 'testicles' – actually were. After squabbling with Van Horne over the male testicle, De Graaf now gave his old teacher credit for being the first to suggest clearly that the ovarian follicles or their contents were eggs (Steno had merely said that the eggs were to be found in the ovaries): 'The distinguished Mr Van Horne, however, pleases to call them *ova*, "eggs" in his "Prodromus" and since this term pleases us better than the rest, we shall use it henceforth as being more apt. Along with that distinguished gentleman we shall call the vesicles "eggs" on account of the exact similarity they bear to the eggs which the ovaries of fowls contain.'[29]

This suggested that De Graaf, like Kerckring, thought the ovarian vesicles (what we now call 'Graafian follicles') actually were human eggs.[30] However, De Graaf's observations, and his attempts to explain them, were more precise and much richer than is generally thought. The key to understanding his vision of generation in general, and of the egg in particular, lies in the final chapter of his book. The experiments described there went far beyond any previous study of the generation of mammals, easily surpassing the intuitions of Steno or the anatomy of Van Horne and Swammerdam. They provided the first piece of real evidence that mammals have eggs, and made De Graaf's book one of the classics in the history of science.

In a series of rigorous experiments that mirrored Harvey's studies of deer in *De Generatione*, De Graaf dissected female animals at various intervals after copulation,[31] noting changes in the colour and number of follicles in the 'testicle'. But rather than following Harvey and studying large, unreliable wild animals that mated once a year, De Graaf deliberately chose rapidly reproducing, small and placid domestic mammals in which he could accurately record the time of mating: rabbits. This excellent choice had a decisive and contradictory effect on his conclusions. De Graaf did not know it, but rabbits

De Graaf's dissection of a human ovary (1672). The 'eggs' (in fact the follicles) are the round structures marked 'B'.

are unusual in that mating induces ovulation; if he saw mating, it was certain that he would subsequently see a change in the shape and colour of the ovarian follicles if he looked closely enough. Although this helped him come to the right general conclusions about the role of follicles, it also severely misled him over the mechanism of generation in humans.

Like Harvey, when De Graaf carried out his animal dissections, he could find no trace of 'anything like semen' in either the uterus or the vagina shortly after mating. However, at six hours after mating, the follicles had begun to redden; over the next two days they swelled up and grew redder still until, at fifty-two hours, they burst, turning into 'a glandulous kind of matter in the middle of which was a small cavity'; over the next few days the follicle gradually changed colour and faded away.[32] Faced with this unexpected observation, De Graaf 'began to suspect that the limpid substance of the follicles . . . had been disrupted or expelled'.[33]

He found the decisive piece of evidence in a rabbit he dissected at

three days after mating: four follicles had ruptured and in the
Fallopian tubes he found four small spherical structures which he
thought were eggs. The fact that the number of these spheres was the
same as the number of empty follicles in the ovaries was telling, but
it was still possible that neither spheres nor follicles had a direct link
with generation. The picture became clearer in the rabbits he dissected after more time had passed since mating: they generally showed
the same relationship between the number of empty follicles and the
number of embryos. Indeed, the greater the interval, the more convincing the evidence, as rather than being 'something like an egg',
the structures he observed in the uterus gradually became evident
embryos. This was astonishing: De Graaf's rabbit dissections indicated that something that came out of the ovaries turned into an
embryo.

However, De Graaf's evidence was not entirely consistent. In the
rabbit that he dissected six days after mating, there were not as many
embryos as there were empty follicles. But rather than questioning
his theory, he concluded that the missing eggs 'came to a sinister
end'.[34] As it happens, he was absolutely right – about one-third of
rabbit embryos spontaneously abort – but the fact that the number of
empty follicles and the number of embryos did not always coincide
could have been used by critics to undermine De Graaf's hypothesis. He shrewdly countered this potential criticism by pointing out
from the opening pages of his book that there were 'never more foetuses in the uterus than follicles in the ovaries'.[35]

From his observations of the rabbit, De Graaf concluded that
mating had induced the release of the eggs from the follicle; furthermore, like Harvey, he had shown that there was no semen to be
found in the uterus after copulation, and yet women (and rabbits)
could become pregnant without ejaculation occurring directly in the
vagina – this occurs either because of the release of small quantities
of sperm prior to ejaculation or, more rarely, when ejaculation takes
place just outside the vaginal entrance, and sperm swim up the
vaginal passage. De Graaf put these facts together and mistakenly
concluded that only 'the finer part' of semen, a kind of 'seminal

vapour',[36] reached the ovaries after ejaculation, fertilised the eggs and caused them to burst out. This had the advantage of explaining how even a small quantity of semen, ejaculated in a very short time, could nevertheless fertilise a large number of eggs in some animals – it was the 'vapour' that was responsible. It was also entirely in keeping with Harvey's suggestion that semen might act like 'contagion' or a bolt of lightning.

De Graaf used this theory to explain the role of the Fallopian tubes. In mammals there was no direct connection between the Fallopian tube and the ovary, making it difficult to see how anything like an egg could get from the ovary into the uterus. By injecting water into the uterus, De Graaf showed that the only route in and out of the top of the uterus was through the Fallopian tubes. He therefore concluded that 'the real function of the tubes is first, in a fruitful act of coitus, to grant a passage to the finer portion of the male semen making for the "testicles" and thus to enable the eggs in the "testicles" to be made wet; second, when the eggs have been fertilised in this way and expelled from the "testicles", to receive them into their extremities and to take them down through their inner cavity to the uterus'.[37] By focusing on the liquid content of the follicles, not their large external structure, De Graaf sidestepped the apparently insoluble problem of how the eggs entered the uterus. As to exactly how they got from the ovary into the Fallopian tube, he had no idea, simply stating that they travelled 'in an inexplicable but nonetheless visible way'.[38] He would no doubt be gratified to know that more than 330 years later, we still do not fully understand how this takes place.[39]

De Graaf made a spectacular mistake when he extrapolated from the rabbit, in which ovulation is induced by mating (although not by the action of semen on the follicles, as he understandably thought), to all animals, including humans. This had very clear consequences for what might be seen in women who had never had sexual intercourse. As De Graaf put it, criticising Kerckring without naming him: 'those who say that both married women and virgins excrete the eggs we have so often mentioned are wrong. The glandulous substance through whose agency the eggs are dislodged from the ovaries begins to grow inside the egg follicles only after a fertilising act of coitus.'[40]

Although this could be read to mean that he thought virgins had eggs but did not release them, elsewhere De Graaf said that the empty follicles in the ovary, one of which is in fact produced each month by ovulation in a fertile women's ovaries, 'do not exist at all times in the female "testicles". They can be detected there only after coitus.'[41] This was plain wrong and could be disproved; dissection of a virgin who had died two to three weeks after her last period would confirm the presence of an empty follicle and therefore the release of an egg some days before. The particular situation in the rabbit made it easy for De Graaf to demonstrate the link between the empty follicle, mating, fertilisation and embryonic development. His evidence would have been much less convincing had he chosen to study a more typical mammal – release of the egg would take place without mating, and it could have been argued that it had no connection with generation.

As appreciated by some of his contemporary readers,[42] De Graaf made a distinction between the follicle – the structure in the ovary that is covered with a membrane and which releases the oocyte – and the egg itself. He argued that once the egg had encountered 'a glandulous substance irradiated by the male semen', it 'bursts out through the nipple visible at the tip of the follicle'[43]– a hole so small that a 'hair-bristle' would only just pass through it.[44] Understandably, De Graaf wanted to know what this tiny egg was like. If he pricked a follicle, it released a thin liquid like the white (albumen) of a chicken egg which 'acquires from being boiled the same colour, flavour and consistency as does the albumen in fowls' eggs' (Kerckring thought that human eggs tasted unpleasant; perhaps the bits of the corpse he was working on were more decomposed than the ones De Graaf cooked).[45]

Although De Graaf gave due credit to Van Horne and Steno, and openly grappled with contradictory ideas from the past, he carefully avoided all mention of Kerckring, despite clearly criticising his work when dismissing the idea that human eggs could grow to be the size of a cherry in three days.[46] But at least this criticism showed that he acknowledged Kerckring's existence; Swammerdam fared far worse: he was not referred to at all, either implicitly or explicitly. Whether accidental or not, this looked like a deliberate snub and must have aggravated the tension between the two men.

Shortly after the publication of De Graaf's work, war broke out. Two kinds of war: a literary and personal war between De Graaf and Swammerdam, and a real, terribly bloody war as the French, supported by their English allies, launched a surprise attack on the Dutch Republic. Over the next year and a half, the two conflicts ran in parallel: the rivalry between De Graaf and Swammerdam grew in ferocity, while the Dutch were caught between the English fleet's attempt to blockade the coastal ports and French troops storming through the southern regions of the Netherlands, bringing terrifying destruction, laying siege to Leiden, and provoking riots and mayhem in Delft and Amsterdam. Both men were understandably disturbed as their stable, successful world – the Golden Age of the Republic – teetered under the stamp of marching feet, the crash of cannon-fire and the screams of massacred civilians. Although the tone of their argument has shocked readers over the centuries, some allowance should perhaps be made for the extremely stressful circumstances under which they were writing.

De Graaf fired the first shot. On 28 March 1672 he sent Swammerdam a pre-publication copy of *De Mulierum Organis Generationi Inservientibus Tractatus Novus*, accompanied by a note challenging his former friend to criticise it in public, if he could. Swammerdam responded by publishing two more of the drawings of the uterus he had made with Van Horne, dedicating the figures to 'the most illustrious and most knowledgeable Royal Society', to which he immediately sent a copy.[47] The content of the drawings showed what he thought was really at stake in his growing conflict with De Graaf – not simply giving Van Horne (and himself) the credit for discovering human 'eggs' but also for describing the anatomy of the whole female genital system. As well as the illustrations of the uterus, the plates included a small diagram of the female hymen (a rather sad ring of ragged tissue) and some striking close-ups of the female external genitalia, showing the structure of the clitoris. One of the figures showed a rather poorly drawn dissected ovary, with what appear to be five pea-sized follicles (Swammerdam explained that the 'eggs' were shown larger than in real life). But this was merely incidental to the elaborate presentation of the uterus and its arteries, which, as in the figure dedicated to Dr Tulp, were hand-coloured in vivid red paint. Once again, this was not a work that dealt primarily with generation, except in the sense that it portrayed the anatomy of one of the key organs involved.

Swammerdam managed to get his material to London just before international communications were affected by the war. When the Royal Society discussed the evidence a few weeks later, in April 1672, they were described as 'concerning the structure of the *uterus* and *ovarium* belonging thereto'. The report of the meeting continues: 'It was recommended to the consideration of Dr. SMITH and Dr. BROWN, who were desired to make a report to the Society at their next meeting, of what might be peculiar in it.'[48] But Drs Smith and Brown, like so many of their Royal Society colleagues given similar tasks, never reported back, nor were they reminded of what they had been asked to do.

At around the same time Swammerdam wrote to De Graaf, reproaching him for the collapse of their friendship, grudgingly acknowledging the accuracy of his experiments on the rabbit and suggesting that he would have done better to present only his own work, leaving other discoveries to the learned men who had made them, 'without attributing them to yourself'. Having accused De Graaf of stealing other people's work — exactly the thing De Graaf had been so concerned about — Swammerdam closed his letter with a cryptic threat, saying he would ensure that the honour for the discovery of the ovary went to those who deserved it.[49]

Exactly what Swammerdam meant became clear a few weeks later, less than two months after the appearance of De Graaf's book, when he published *Miraculum Naturae, sive Uteri Muliebris Fabrica* ('The miracle of nature, or the structure of the female uterus'), two hundred pages long, which eventually sold in London for 2s 6d – half the price of De Graaf's book.[50] A long secondary subtitle explained that the book consisted of Swammerdam's detailed comments upon Van Horne's 1668 *prodromus*. But paradoxically, the *prodromus* dealt mainly with the male genitalia, not with the female organs. In both cases, Swammerdam clearly felt he had to defend Van Horne's honour against De Graaf. He therefore mounted a double attack on his one-time friend.

Swammerdam's book was obviously written quickly, using Van Horne's *prodromus* and De Graaf's *De Mulierum Organis Generationi Inservientibus Tractatus Novus* as twin templates. But unlike De Graaf, Swammerdam had not carried out a systematic study of the anatomical origins of generation in females. He had made some relatively straightforward dissections of the human female genital organs with Van Horne, and that was about it. He therefore filled his pages with information that was interesting but not to the point, such as a description of his wax-injection techniques, criticisms of other anatomists' views of the function of the diaphragm and even a long discussion of Malpighi's dissection of the silkworm brain. Despite its title, less than half of Swammerdam's book dealt with generation or the structure and function of the female genitalia.

In his most important criticism of De Graaf's view of generation, Swammerdam used his own observations and those of the Hague physician and anatomist Stalpaert vander Wiel to insist, correctly, that 'cavities and traces' could indeed be seen in the ovaries of virgins, showing that they had released eggs. Swammerdam also addressed the 'grave and difficult' problem implied by the radical new description of the role of women – how the eggs moved from the ovaries to the uterus. Like De Graaf, Swammerdam thought they passed through the Fallopian tubes, but did not see how 'the matter contained in an egg' could get through them.[51] Even if, as Steno had suggested, and Swammerdam agreed, it was the fluid in the follicle that was really the egg, in dissected women the Fallopian tube generally appeared to be about 'three fingers'-width' from the ovary, in dead women at least. Swammerdam felt this was too far even for a fluid egg to travel. He put forward two solutions to this difficulty – some kind of peristaltic motion within the tube might transport the egg towards the uterus, and perhaps the distance between ovary and tube was not so great in the living body. Both of these explanations are at least partly true.

Apart from all the asides about silkworms and wax injections, and the criticisms of De Graaf's description of the male genitalia, the central theme of Swammerdam's book was that Van Horne was the first person to describe accurately the male and female genitalia, and to suggest that women have eggs. Swammerdam also repeated his claim that he had played a key role in Van Horne's work, stating that the three illustrations in *Miraculum Naturae* (composed of the sheet dedicated to Dr Tulp, and the two figures sent to the Royal Society earlier in the year) had all been made at Van Horne's request in 1667. To back this up, Swammerdam listed the names of the people who had seen the drawings at the time – including De Graaf, who, interestingly, never disputed this. Given De Graaf's readiness to pick up on the slightest mistake, this can be taken as an admission that he had indeed seen Swammerdam's drawings. Swammerdam also somewhat inconsistently declared that he was not 'seeking to claim for myself the honour due to another'[52] and that Van Horne was the person who deserved the credit.

All this would have been enough to irritate De Graaf, but he was sent over the edge by two further parts of Swammerdam's book. Firstly, Swammerdam now claimed that it was he who had suggested to Van Horne that the egg was in the follicle and that it went down the Fallopian tubes.[53] The unstated implication was that the true credit for the discovery of eggs should go to Swammerdam. Whatever the truth – and we will never know whether Swammerdam was lying or not – De Graaf was infuriated. Second, *Miraculum Naturae* concluded with a brief appendix which contained no new information but plenty of allegations and abuse. In less than four pages, Swammerdam sarcastically criticised De Graaf for not mentioning Kerckring, and used some minor errors to cast doubt on the accuracy of all of De Graaf's drawings. He then turned the tables by suggesting that De Graaf had been inspired by the drawings he had done for Van Horne, described his former friend as 'conceited', before boasting that his *Miraculum Naturae* contained no dissections 'of rabbits and brute animals', which were inevitably 'inferior' to investigations of humans.

Swammerdam dedicated his book to the Royal Society, sent a copy to London and asked the Society to decide who was right in all the disputed questions of anatomy and priority.[54] Two months later, to back up his claims, he sent a beautifully preserved female uterus, together with twelve other items of genital anatomy, including a dissected penis, a clitoris and a hymen. According to the catalogue of the Royal Society's museum, the uterus was 'prepared after the method of Dr. SWAMMERDAM, with all the other parts dried up, and the vessels filled with yellow and red wax, very distinctly injected'.[55] When the specimens eventually arrived in London in December 1672, 'after a checkered career and countless perils of war', as Oldenburg put it,[56] they became a prized part of the Society's collections. Oldenburg, who thought the bits of body were 'very fascinating and prepared with exceeding ingenuity',[57] apparently became rather attached to them: several years later he was upbraided by the other Fellows of the Royal Society for taking them home for too long.[58] The specimens eventually found their way to the British Museum, before disappearing from sight at the beginning of the nineteenth century. The uterus may still be lurking somewhere in Bloomsbury, but it was probably discarded many years ago – even at the time Swammerdam warned that it could become mouldy and would need regular brushing with 'oil of turpentine'.[59]

While Swammerdam was busy writing his attack on De Graaf, what the British call the third Anglo–Dutch war had begun. In April 1672, Louis XIV activated the secret Treaty of Dover, which he had signed

in 1670 with his cousin, Charles II. In a rather one-sided deal, the
two monarchs had agreed to attack France's economic rivals and
political and religious enemies to the north. There was little to be
gained by the English and, when the war began, few people outside
Charles's court seemed to feel it was a war worth fighting. In the initial stages, however, the war had the huge advantage of being
successful. The Dutch suspected nothing, and were completely
unprepared for the lightning French attack. The enemy troops swept
northwards, bringing havoc and destruction, helped by the severe
drought of the preceding months, which had left the canals unusually low – in previous conflicts the Dutch had cut off advancing
armies by the costly but effective tactic of flooding the land. Within
two months, the invaders had seized Utrecht, and the political leaders of the Republic were under severe pressure. In June and July,
Delft was shaken by violent rioting. The calm streets portrayed by
Vermeer were full of protesting crowds, fighting with the militia and
calling for action by the government as the French army drew closer.
Everywhere, people felt that the Dutch Golden Age was threatened,
and might even be coming to an end.[60]

At the same time as the French troops were bringing death and
destruction to nearby cities, De Graaf married Maria van Dijk in
Delft, in June 1672. In a practical example of the wonders of generation, Maria immediately fell pregnant, adding to his worries. In a
confused letter written to Oldenburg in July, he apologised for not
writing earlier, explaining that 'the disaster falling upon the whole of
my country stifled all desire for correspondence'.[61] The letter, which
is in the Royal Society archives, revealed De Graaf's agitation. Hastily
written, full of crossings-out and with several smudges, its appearance
was in such sharp contrast to the ordered and neat presentation of his
previous letters to the Society that he felt obliged to add a pathetic
postscript – 'Forgive my disordered and hurried handwriting.' De
Graaf was clearly not a happy man.

Over the next few months the Dutch political situation went from
turmoil to crisis. In July, following a series of riots against a government that seemed unable to stop the French advance, Prince William

of Orange, the hereditary monarch and future King William III of England, seized power. Two months later an enraged mob in The Hague seized two of the key deposed leaders of the Republic (the De Witt brothers) and tore them apart.[62] This horrific event, and the mass riots which characterised the summer of 1672, left a wound on the Dutch popular and political consciousness which still smarts. In the autumn, the political consequences of William's coup became clear. Orangeist demonstrations in Delft and Amsterdam were followed by a series of purges of all state appointments and of local government officials, as both Catholics and followers of the old political leaders were chased from office – in the state of Holland, William replaced nearly a quarter of the regents. Even small places such as De Graaf's native town of Schoonhaven were affected – amongst the purged may well have been some of the local councillors to whom he had dedicated his doctoral thesis.

Meanwhile, by autumn 1672, more than half of the Republic was in the hands of the French invaders. In the following months, Catholic masses were celebrated in Utrecht Cathedral, and the symbols of Dutch Protestantism were destroyed in the occupied territories. Political, military and economic confidence collapsed. There is no record of either De Graaf or Swammerdam doing any scientific work in this period. Both men were presumably deeply disturbed by the political and military situation, while Swammerdam may have been personally affected by one of the most notorious events in the war. In the winter of 1672–3, French troops carried out horrific massacres in two small villages – Bodegraven and Swammerdam, where his family came from.[63]

De Graaf received a further blow in November 1672, when he learnt of the death of Sylvius. Sylvius, who was buried with full academic honours in the Pieterskerk in Leiden, was the best-loved of all the professors in that city, and his relations with De Graaf had always been very close. Furthermore, if De Graaf had ever dreamt of taking Sylvius's chair at Leiden University, despite the traditional ban on Catholics in university professorships, the reactionary political climate under William of Orange made that completely impossible.

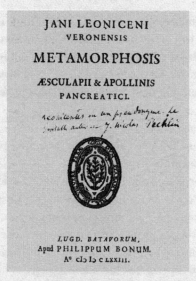

JANI LEONICENI
VERONENSIS

METAMORPHOSIS

ÆSCULAPII & APOLLINIS
PANCREATICI

*reconicentes ou un pseudonyme. Le
véritable auteur est J. Nicolas Pechlin*

LUGD. BATAVORUM,
Apud PHILIPPUM BONUM.
Aº cIɔ Iɔ c LXXIII.

*Title page of the satirical book published in 1673 by 'Leonicenus',
attacking Sylvius and De Graaf. A previous owner has written (in French):
'Leonicenus is a pseudonym. The real author is J. Nicolas Pechlin.'*

In February 1673, De Graaf was extremely preoccupied. He was
working on a response to Swammerdam's accusations in *Miraculum
Naturae*, and his wife was approaching the end of her pregnancy.
Then he found himself ridiculed in the pages of a viciously satirical
book that targeted Sylvius and his medical theories about fermentation: *Metamorphosis Aesculapii et Apollinis Pancreatici* ('The
metamorphosis of Aesculapius and the pancreatic Apollo' –
Aesculapius was the Greek god of healing, and the son of Apollo).[64]
Even the title poked fun at De Graaf's pretensions – the 1671 edition
of his book on the pancreas, which also contained his *prodromus* on
the female genitalia, was prefaced by a poem singing his praises as the
'true Apollo' of the pancreas. What particularly irked him, however,
was that the book also contained a series of personal attacks against
Sylvius, even though he was not long dead. Although the author
used the pseudonym 'Janus Leonicenus of Verona', De Graaf rightly

detected the hand of Johann Pechlin (1646–1706), another Leiden graduate who had written the preface to Swammerdam's 1667 thesis.[65] It would have taken a strong mind not to suspect a plot. In March, Reinier and Maria de Graaf's son, Frederick, was born in Delft and christened in the Catholic church next to Vermeer's house. But within a month the baby was dead. When De Graaf's reply to Swammerdam was published a few weeks later, under the title *Partium Genitalium Defensio* ('Defence of the genital parts'), it was as intemperate as might be expected.

De Graaf's book contained no new scientific evidence; instead, most of the material was strongly polemical. He described the long gestation of the priority dispute, reproducing letters, quoting conversations and using terms that were far more violent even than those Swammerdam had employed in *Miraculum Naturae*. He accused Swammerdam of copying his illustrations, sneered at his elegant wax-injection method, claimed his one-time friend was 'blinded by anger and hatred' and suggested that he piled 'lie upon lie' – all this in the first few pages. Oddly, De Graaf flatly denied that he knew Steno had written anything about eggs, despite having seen him in Delft and corresponded on the question several times. And he scoffed at Swammerdam's claim to have been the first to suggest that eggs descend to the uterus via the Fallopian tubes, although this necessarily remained in the realm of polemic, not proof.

The strangest thing about De Graaf's book was that it was dedicated to Louis XIV, the leader of the attack that was bringing the Dutch Republic to its knees. This could indicate that, as a Catholic, De Graaf was favourable to Louis' cause, or even that he cynically thought such a dedication would serve him, in the event of a French triumph. It seems more likely that, although he was clearly upset and worried by the invasion and its effects on Dutch society, he did not view everything through the prism of the war. Other people took equally open attitudes. Shortly before he died, Sylvius had been contacted by a member of the French royal family for a medical consultation; Schacht mentioned this with pride during his funeral oration to his dead colleague, at a time when the French troops were nearly at the gates of Leiden.[66] And after the outbreak of the war, Oldenburg wrote to Swammerdam that 'It is indeed proper that honest and true philosophers should, while the princes of the world contend fiercely over questions of *mine* and *thine*, persist in the peaceful search into nature's secrets.'[67]

As its name suggests, most of *Partium Genitalium Defensio* is a defence of De Graaf's version of male and female genital anatomy, against Swammerdam and Van Horne. It had no direct consequence on the key problem of generation, although it added fuel to the polemical fire. The language De Graaf used surpassed even the bad-tempered tone of his previous literary conflicts. In a final outburst, he roundly abused Pechlin, saying 'your book came out of your arse not your head'.[68] He then dedicated this understandably unbalanced and furious book – he himself called it 'not very polite' – to the Royal Society and, like his rival, asked them to decide on the points that separated himself and Swammerdam.

In June 1673, the Royal Society received De Graaf's book and asked three of its members – Drs Needham, Croune and King – 'to give the society an account of it'. As always in such circumstances, it took longer for the Fellows to report back than was initially expected. But, unusually, within a few weeks there were indications that progress was being made. The informal chair of the group, the 62-year-old Needham, reported: 'as to the dispute about the said discovery of eggs in viviparous animals, the readers must be referred to the times, when the several claimants of that discovery published their books about it, and thence left to judge of the priority contended for'.[69] In other words, Needham did not intend to get involved in the debate over priority and wanted to restrict the committee's work to the anatomical detail. And even on this point, he sat tightly on the fence: 'in some things DR DE GRAAF was in the right, and mistaken in others, and *vice versa* DR SWAMMERDAM'.

By August the three men had come to an agreement and had written their ten-page report. But as the Royal Society did not meet in the summer, the papers remained with Oldenburg for six more weeks until that wet morning in October when the Council met in Lord Brouncker's office. Surprisingly, the verdict was not as Needham had initially expected. From the outset, the three men came to a very clear opinion about priority. But it was not one that either Swammerdam or De Graaf would have wished for. The report stated that the first person to have seen eggs in viviparous organisms was not Van Horne, nor De Graaf, nor even Swammerdam, but Steno, as reported in his 1667 dogfish dissection.

This was an acute move. Firstly, it stuck to the established principle that priority went to those who published first.[70] Retrospective claims, no matter how justified, were less important than actually getting into print. Second, this decision made an important scientific statement, by lumping together all viviparous animals, including humans. Rather than dealing simply with human eggs, the Royal Society was underlining the existence of the new vision of generation in all animals. Finally, by playing down the specific importance of human eggs, it effectively sidestepped the ferocious priority debate between Swammerdam and De Graaf, which immediately appeared minor, and even petty.

Of the report's nine points,[71] only three dealt with the female genitalia, and these were each dismissed in a couple of sentences. Either because the report's authors thought that the male genitalia were more important, or because they felt the questions raised by the rivals' claims were more thorny with regard to male structures; most of the report dealt with the detail of which vein goes where and which nerve feeds which part of the penis. Swammerdam and De Graaf had invested a huge amount of time and effort in their dispute over the egg, only for the Royal Society largely to ignore the question.

Tragically, about a week before the committee finished its report, De Graaf died, aged only thirty-two.[72] He was buried in the Oude Kerk in Delft, next to his baby son; his wife Maria disappears from history. There is no record of what caused De Graaf's death, but more than twenty years later Leeuwenhoek recalled that he had heard at the time that Swammerdam and De Graaf 'had such a sharp verbal altercation that not only did the latter fall ill, but this was also followed by his death'.[73] Whatever the truth – it seems likely we will never know what happened – there was no mention of De Graaf's demise in the minutes of the Royal Society, or in the letters of any of its correspondents. Over a month later, when it discussed the report, the news had still not arrived.

As for Swammerdam, who had been busy working on generation in bees, he received the report and replied to it almost immediately. On 7 November 1673, at the second meeting of the Society to take place that day, 'Mr Oldenburg began to read Dr. Swammerdam's Answer to the Letter formerly sent to him by three of the Physicians of the Society'. However, 'There being not time enough to make an End of it at this Meeting, the reading of the Remainder was referred to the next.'[74] Frustratingly, nothing more was heard of the letter. In an uncharacteristic slip, Oldenburg did not enter it into the Royal Society's Letter Book; no trace of it or nor even the slightest hint as to its contents can be found in the archives. Swammerdam never mentioned De Graaf or the dispute again, although he did continue – quite rightly – to claim to have observed human 'eggs' with Van Horne in 1667.[75]

The unwitting victor, Steno, was on the verge of abandoning science altogether, and there is no evidence that he ever heard what had been decided; nor perhaps would he have cared. From the outset, he had been happy to give priority to Van Horne, and never made any great claims over his discovery. Even the Royal Society seems to have lost heart over the question: the report languished in the archives for more than eighty years, before being published, in the original Latin, as a few pages in a four-volume collection of the minutes and discussions of the early years of the Society.[76]

This unflattering episode has been the subject of many comments over the centuries. At the time, readers were generally uncritical: the review of Swammerdam's *Miraculum Naturae* in the *Philosophical Transactions* highlighted the key differences with De Graaf over various points of anatomy, but did not suggest there was anything

improper about the tone or the claims he made,[77] while the authors
of the Royal Society's report ignored the vocabulary used in both
books. Isbrand van Diemerbroeck, who had been De Graaf's teacher
at Utrecht, was more partial. In 1674 he wrote that in the pages of
Miraculum Naturae Swammerdam seemed 'to besmear the whole
Ovary together with the Eggs, not with *Honey*, but with the most
bitter *Gall*'.[78] Later writers have also tended to take De Graaf's side:
at the end of the eighteenth century Antoine Portal, one of the first
historians of anatomy and surgery, said Swammerdam's behaviour
was 'very bizarre',[79] while one twentieth-century historian described
his claim to have observed the 'egg' with Van Horne before De
Graaf as 'a revolting pretension'.[80]

These criticisms seem unfair. In terms of the tone, De Graaf gave
at least as good as he got, and he had a track record for tetchy defen-
siveness, as shown by his polemic with Clarke over the testicle and
his fears expressed in both his 'advance notices'. Most importantly,
however, the kind of language both sides used was typical of the
time. Neither Swammerdam nor De Graaf was a genteel eighteenth-
century amateur, nor were they modern scientists constrained in a
straitjacket of acceptable scientific language. They were pioneers
whose determined defence of their ideas was exacerbated by a
friendship gone sour.

As far as the justification of each man's claims is concerned,
Swammerdam's suggestion that he had observed and identified the
human egg (or, at least, the ovarian follicle) before De Graaf is
incontestable, although whether either man profited from seeing the
other's figures prior to publication is impossible to determine. More
significantly, we can be reasonably sure that none of the participants
in this dispute – Van Horne, Swammerdam, Kerckring, De Graaf or
Steno – actually saw the human egg prior to fertilisation.[81] If they
did see it, they did not leave an unambiguous description which we
can recognise. A fair balance sheet would underline how each par-
ticipant – not just Swammerdam and De Graaf – helped create the
widespread acceptance of the egg hypothesis. Van Horne first spelt
out the idea that humans have eggs in ovaries and suggested how the

egg gets to the uterus, Steno first stated this in print and generalised the idea to all viviparous organisms, Swammerdam first illustrated the ovaries and the follicles in clear detail, while De Graaf provided the first experimental evidence that eggs are released from the ovary. Strangely enough, Kerckring, whose scientific contribution was virtually non-existent, had by far the greatest immediate impact on contemporary views – in the 1670s and 1680s, many non-specialist readers were more convinced by Kerckring's silly jelly babies than they were by De Graaf's precise studies of rabbit reproduction or Swammerdam's delicate drawings.[82]

There were three reasons for this anticlimactic conclusion to the ferocious priority dispute. Firstly, the egg hypothesis was so successful that those who accepted it did so without differentiating between the various proponents – it was only in the eighteenth century that De Graaf began to gain the credit he deserved, when the term 'Graafian follicles' was adopted to describe the ovarian vesicles. Second, the death of De Graaf effectively defused the row; this was reinforced by the fact that for the next three years Swammerdam became increasingly obsessed with the strange religious ideas of the itinerant French preacher Antoinette Bourignon, to the extent that for a while in 1675 and 1676 he abandoned science altogether.[83] The final, most important reason was that the debate soon appeared completely outdated. Events were moving so quickly that within five years of the widespread acceptance of the egg hypothesis it was being fiercely attacked by one of De Graaf's friends from Delft.

MAN COMES NOT FROM AN EGG

It cannot have felt that way at the time, but science hit a barrier in 1672. Although enormous progress in understanding generation had been made through the power of ideas and the use of experimentation and careful dissection, De Graaf's work on the egg marked the limits of knowledge that could be obtained with existing techniques. However, the scientific revolution was not just about doing experiments; it also involved using instruments to study the natural world, revealing structures and processes that were previously impossible to investigate or even to imagine. The next step on the road to understanding generation required a shift in perception as radical as that which occurred when Galileo turned his telescope to the heavens at the beginning of the century.

In April 1673, four months before his death, Reinier de Graaf sent his last scribbled letter to the Royal Society, accompanying a copy of *Partium Genitalium Defensio* and some papers written by one of his acquaintances from Delft. De Graaf wrote: 'So that it may be clear to you that humane and philosophical studies are not yet banished from this place by the din of war, I will communicate to you at this present time what a certain very ingenious person named

Leeuwenhoek has achieved by means of microscopes.'[1] Unlike De Graaf, Swammerdam, Redi, Steno and Malpighi, Antoni Leeuwenhoek (pronounced 'Lay-wen-hook') was not a physician, nor even a young student. Aged forty-one when De Graaf's letter was sent, he sold cloth, buttons and ribbon in his draper's shop. He had no academic training and he could not speak another language, not even Latin; all his correspondence with foreigners had to be translated. Despite this, he became one of the Royal Society's most productive members, astonishing the world with his discoveries and attracting visitors from all over Europe to his small house in Delft to peer through his tiny home-made microscopes.[2]

The microscope was invented in the Netherlands around the turn of the seventeenth century, but was initially seen as little more than a toy – the earliest models were barely more powerful than magnifying glasses. As with the telescope, the full potential of the new instrument was first realised in Italy, where the term *microscopio* was coined.[3] In 1625 Francesco Stelluti (1577–1652), who, like Galileo, was a member of the Accademia dei Lincei, published the first illustration using a microscope – a drawing of three huge honeybees, set out in a pastiche of the Pope's coat of arms.[4] Using copper engravings, Stelluti's broadsheet revealed bees as hairy, armoured organisms with a complex anatomy to rival that of larger animals. Compared to previous rather crude woodcut illustrations of insects, these creatures came from another world.

Stelluti's innovation had very little immediate influence, partly because of the death of Prince Cesi, who had founded and bankrolled the Accademia dei Lincei, and partly because of the Church's campaign against the most famous member of the Lincei – Galileo. Nevertheless, over the next forty years a number of brief studies using the microscope were published in Italy, including descriptions of a fly's eye, of the eyes of a spider and a series of descriptions of insects.[5] But the impact of these scattered studies was limited by the poor performance of the microscopes and by the fact that the observations were presented in a few low-resolution woodcuts.

The great change in attitudes towards the microscope began in England, as the country emerged from the Civil War and launched itself into the revolutionary experiment of Cromwell's Commonwealth. John Wilkins, the Master of Wadham College, Oxford, was particularly interested in the microscope and the 'strange discoveries of extream minute bodies' he thought it would permit.[6] In the late 1640s Wilkins organised a group of thinkers to investigate the natural world; this circle was one of the component parts of the future Royal Society and included the young Christopher Wren, who was Wilkins' student. Wren had worked out how to draw the images that could be seen under the microscope, and in the late 1650s he produced incredibly detailed pictures of a flea, a louse and a mite.[7] After the restoration of the monarchy in 1660 these were shown to the newly crowned Charles II, who was most impressed and ordered Wren to make some more. But despite being given help by the Royal Society, Wren made no progress.[8] Luckily for the ambitious young man, Charles also lost interest as more exciting diversions attracted his attention, and the royal command appears to have been forgotten.

Although Wren abandoned his studies with the microscope, his friend Robert Hooke did not. At the beginning of 1663, Hooke, who had recently been appointed Curator of Experiments at the Royal Society, was urged by the Society 'to prosecute his microscopical observations, in order to publish them' and 'to bring at every meeting one microscopical observation at least'.[9] Over the next few months he diligently produced drawings of objects as varied as a razor, a ribbon, a millipede and a gnat, eventually finishing his work in June 1664. The book appeared six months later – Lord Brouncker had first to check it would not harm the good name of the Royal Society.[10]

Micrographia, or Some Physiological Descriptions of Minute Bodies Made by Magnifying Glasses with Observations and Inquiries Thereupon, to give Hooke's book its full title, was an instant and hugely influential success. Pepys saw the unbound printed sheets in a bookshop and immediately decided to buy a copy. When he eventually got his

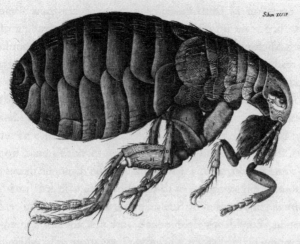

Hooke's drawing of a flea from Micrographia. *The original is about 50cm across.*

Hooke's drawing of 'blue mould found on brown leather' from Micrographia.

hands on it, on 21 January 1665, he wrote in his diary: 'Before I went to bed, I sat up till 2 a-clock in my chamber, reading of Mr Hooke's Microscopical Observations, the most ingenious book that ever I read in my life.'[11] The Dutch astronomer Christiaan Huygens had to tear himself away from reading it in order to write a letter to London congratulating the Royal Society on the amazing quality of the observations.[12] Pepys and Huygens were right. Hooke's book is simply astonishing.[13] It contains sixty observations of a variety of objects, from a needle to a full stop, from a spark to a gnat, from an ant to some mould, mainly describing what he could see through his 'compound' microscope – a 15-cm tube with a lens at each end, very similar to that used by Stelluti at the beginning of the century. Written in a confident style, Hooke's work describes his findings and explains how he made them, including suggestions as to where readers could buy a microscope like his.

What truly seized the public imagination were the illustrations. These still retain their power: the thirty-eight minutely detailed figures reveal both Hooke's skill as an artist and the incredible detail that could be shown with the fine lines of a copper engraving. Some of them were nightmarish, such as the giant fold-out illustrations of a flea and a louse – Christiaan Huygens said each was 'as big as a cat'[14] – while others had a strange, dream-like quality, such as the drawing of blue mould. Although only around one-fifth of Hooke's observations described organic objects, these were the most stunning illustrations in the book, revealing an unimaginable beauty, complexity and downright weirdness in the living world.

In retrospect, Hooke's most important discovery went unnoticed at the time: he described how a slice of cork – the dead bark of a tree – was composed of small, separate empty units, which he called 'cells' after the rooms in a monastery. It would be more than 170 years before the full importance of this observation became clear – all life is made of cells. Hooke may have seen only the dead spaces which once held the living, interacting cells that made up the plant, but the impact of his casual analogy was enormous: every day millions of people around the world now use the word chosen by Hooke nearly 350 years ago, to refer to the fundamental unit of life.

Micrographia was so influential, and encouraged so many people to look down a microscope, that it has been described as the first popular science book, designed as much to inspire and educate a lay readership as to inform fellow scientists.[15] On the Continent, the fantastic illustrations played an even more important role, as the book was written in a barbaric language that few could understand – English.[16] However, despite the immediate attraction and impact of Hooke's work, microscopy was notoriously difficult, as many rich readers of *Micrographia* found to their cost. Pepys had earlier bought a microscope from Reeves, the dealer recommended by Hooke, for the huge sum of £5 10s but was disappointed with the results, being unable to see anything like the hyper-real clarity of Hooke's illustrations.[17]

Furthermore, not everyone agreed that Hooke's observations actually helped in understanding the world. In 1666 Margaret Cavendish, the Duchess of Newcastle, who had herself used a microscope in the 1640s,[18] scornfully dismissed 'the lately discovered Art of Micrography' as simply describing the outside of tiny animals, pointing out that the technique was unable to 'discover their interior, corporeal, figurative motions, and the obscure actions of Nature, or the causes which make such or such Creatures'.[19] Although not many thinkers agreed with the Duchess, even this partially justified criticism of Hooke's work was soon swept away by the torrent of microscopical studies which poured out of the Netherlands.

The man who did most to show the power of the microscope, Antoni Leeuwenhoek, was not only a draper, he was also 'Chamberlain of the Council-Chamber of the Worshipful Sheriffs of Delft', a qualified surveyor and a man of some means who owned 'gardens' both inside and outside the Delft city walls.[20] In September 1676, shortly after becoming a regular correspondent of the Royal Society, he was made 'curator of the estate and of the assets of Catharina Bolnes, widow of the late Johannes Vermeer during his lifetime master painter', after she

Portrait of Antoni Leeuwenhoek, by Jan Verkolje.

had petitioned to be made bankrupt. While this might imply that he was a friend of the Vermeer family, there is no record of any link between the two men in Vermeer's lifetime. Whatever the case, Leeuwenhoek did the family no favours, blocking Catharina's attempts to transfer some of her assets to her mother by seizing and selling Vermeer's masterpiece *The Art of Painting*, despite the family's opposition.[21] Leeuwenhoek's profile and position were very different from those of most participants in the scientific revolution, virtually all of whom either did not have jobs or were professionals and who, in general, were closely connected to the aristocracy. It is hardly surprising that the Dutch diplomat Constantijn Huygens described Leeuwenhoek, with a mixture of affection and condescension, as 'our bourgeois philosopher'.[22]

No one knows why Leeuwenhoek began to use the microscope.

Perhaps he wanted to study the weave of the cloth he was selling, or perhaps he was inspired by *Micrographia*, which he might have come across during a visit to England in the mid-1660s, or perhaps he simply enjoyed using his undoubted ingenuity and manual skills.[23] Behind these explanations – all of which might be true – lies a general assumption that there was something in the atmosphere of the Dutch Republic which led scientists and artists, and in particular Leeuwenhoek and Vermeer, to use optical devices to portray the world.[24]

The letter De Graaf parcelled up in the spring of 1673 contained the first record of Leeuwenhoek's observations with a microscope. Consisting of a description of a bee sting, a louse and a moss, Leeuwenhoek's brief report was published by Henry Oldenburg in the *Philosophical Transactions* that October. It was the first of 190 letters from Leeuwenhoek to the Royal Society, sent over a period of more than forty years, most of which were published. In recognition of the importance of his contributions, he was made a Fellow of the Royal Society in January 1680. At the time when Swammerdam and De Graaf were engaged in their polemical squabble and De Graaf at least was deeply disturbed by the political and military crisis that gripped the Dutch Republic, Leeuwenhoek was apparently unaffected by events, carrying out the investigations he eventually sent to London. Although the third Anglo-Dutch war petered out in 1674, with the English gaining virtually nothing (except New York, which the Dutch had retaken during the conflict), Louis XIV's offensive against the Republic continued until 1678. Unlike De Graaf and Swammerdam, Leeuwenhoek appears to have sailed through the war and the crisis it created with no discernible ill effects. However, this may be a false impression, a trick of the light due to the incomplete archival traces of Leeuwenhoek's life and opinions; he certainly followed political events, as shown by his references to the English 'Popish Plot' crisis of 1688–9, during which the President of the Royal Society, Sir Joseph Williamson, was thrown into the Tower of London.[25]

Although Leeuwenhoek's early studies were basically an extension of Hooke's observations of insects and other organic objects, his technique differed in one critical respect. Like many Dutchmen, he used a microscope which was much more powerful than Hooke's two-lens compound instrument. Sometime in the late 1650s, the Dutch had realised that extremely powerful lenses could be made out of very tiny balls of glass, each around 1mm in diameter. The single-lens microscope revealed things that were far smaller than any described by Hooke; with a typical magnification of around 150x, and perhaps even up to 500x, it allowed the skilled observer to discern

One of Leeuwenhoek's microscopes. The lens was placed in the tiny hole in the top plate. The object to be viewed was put on the spike just next to the hole, and could be moved using the screw and clamp. The whole apparatus is about 8cm long.

objects that were around one micron (a thousandth of a millimetre) in size.[26]

The single-lens microscope was simple and cheap to make; it was merely a wooden or metal frame that held the tiny glass sphere (made by drawing out a thin rod of red-hot glass into a small blob, then shining it with jeweller's polish), plus an attachment for holding the object. But it was far from simple to use, threatening the viewer with a nasty case of eyestrain as he brought the minute lens very close, generally in bright sunlight. The slightest movement of the object would make it move out of the narrow depth of field that was the optical price for having such high magnification. This meant that, despite the small size of the microscope, it was not a portable device in the true sense of the term: it was very hard to make observations away from a desk or bench. Leeuwenhoek told the Royal Society that he found it difficult to study samples if they had to be examined immediately after being collected because he needed his 'customary seat and instruments' to use his tiny microscope.[27]

While these problems beset everyone who tried to use a microscope, Leeuwenhoek had a double difficulty. For people like Wren, Hooke and Swammerdam, who were as skilled with the pen as they were with the microscope, 'all' they had to do was to use their hand to represent what their eye saw. (Anyone who has tried drawing what they can see through a microscope will know that this is in fact extraordinarily difficult.) But Leeuwenhoek, as he recognised, could not draw for toffee; he therefore had to employ a series of artists in Delft to make his illustrations.[28] So not only did the Delft draper have to deal with the object going in and out of focus, his artist had to grapple with the problem, too: Leeuwenhoek would have had to regulate precisely the tiny microscope, then hand the instrument to the artist, and describe what he wanted him (or her) to see. But the artist's eyes would have seen differently from Leeuwenhoek's, the object might not be in focus for him, or it might have moved, or – most likely – he would not be able to see exactly what Leeuwenhoek wanted to show. An extraordinary amount of effort and frustrating misunderstanding must have been involved in making these illustrations.

When Leeuwenhoek's first letter arrived from Delft, Oldenburg and his colleagues were pleased to hear that someone else was using the microscope – Hooke had stopped making observations some nine years earlier, because of eyestrain and the pressure of his other interests. But Oldenburg was well aware that Leeuwenhoek's report did not represent any decisive advance on the kind of descriptions found in *Micrographia* – the Dutchman was simply looking at everyday objects from the outside and showing how odd they seemed when you got really close up. Sensing that more fundamental things remained to be discovered, Oldenburg encouraged Leeuwenhoek to pursue his investigations, and to study other substances, including bodily fluids such as blood.[29] Leeuwenhoek followed the advice and over the next few years he sent a series of letters to London which were regularly published in the *Philosophical Transactions*. Looking at blood, he saw 'globules' – red blood cells – while in plants he noticed 'globules' moving in their 'vessels'. Even a dried optic nerve revealed itself to be composed of 'globules' which Leeuwenhoek thought might transmit the impression of light from the eye to the brain. Every bit of life, it appeared, contained these 'globules', although neither Leeuwenhoek nor anyone else tried to come to any general conclusion about their role.

All of the letters were written in Leeuwenhoek's typically unsophisticated style.[30] He used common, everyday words to describe what he saw, not Latin terms, and generally wrote in a rambling, sometimes incoherent way which apparently reflected somewhat disorganised thinking. Worse, as he cheerfully accepted, he was extremely pigheaded and did not like being contradicted. This made relations with him sometimes difficult; in a letter to Thévenot, Swammerdam complained: 'It is impossible to go into discussion with him, as he is biased, and reasons in a very barbaric way.'[31]

Two of the reasons for this prickliness may have been Leeuwenhoek's awareness of his inferior social status compared to most of the thinkers with whom he corresponded, and the consequences of his lack of academic – or linguistic – training. He was a shopkeeper, not an aristocrat or a physician, and nor did he enjoy the

protection of a powerful patron. His lack of training may explain the
fact that, unlike most scientific investigators of the period, and in
striking contrast to Swammerdam, De Graaf, Redi, Steno and
Malpighi, Leeuwenhoek never collected his ideas and findings into
a book, with all the obligation to develop a long argument that
would have required, despite being pressured to do so.[32] Instead, he
restricted himself to incredibly rich, but often confused and uncoordinated descriptions of his findings in a massive number of letters,
which he sent out from his small house in Delft to his growing network of correspondents around Europe, but first and foremost to the
Royal Society.

News of some fascinating discoveries by Leeuwenhoek arrived in
London in autumn 1676, contained in a long letter describing the
results of his studies of water, which had continued throughout the
previous year.[33] The first set of observations, made in spring 1675,
showed minute creatures which were smaller than the water flea
Daphnia (described by Swammerdam in *Historia Insectorum Generalis*
in 1669). Although the existence of these creatures was surprising,
they were still just visible to the naked eye, being around half the size
of the dot on this i. Leeuwenhoek's true breakthrough came in
spring 1676 when he tried to discover why pepper is hot. This was
related to his previous attempt to find out why sugar and salt taste
different (because of the shape of the crystals, he argued), but also
had a distinctly applied aspect – spices, which were still an important
part of the Dutch economy, had to be imported at great cost from
the Far East; although Leeuwenhoek did not spell it out, understanding why pepper was hot might have led to the discovery of
cheaper substitutes.

Unsurprisingly, Leeuwenhoek did not find out why pepper is
hot.[34] What he did see was far more exciting: four kinds of tiny
animal, far smaller than could be seen with the naked eye, the
smallest being around one-tenth the length of the eye of a louse, as
he put it. These 'animals' were in fact protists (tiny single-celled
organisms) and bacteria, and Leeuwenhoek was the first person to
see them. And he did not observe just one or two tiny blobs; he

could see thousands and thousands of them teeming in a drop of 'pepper-water' (he had crushed pepper grains in water to make the pepper easier to study). He was so surprised that he repeated his observations over and over again between April and September that year. Virtually every time, he saw tens of thousands of tiny animals, and even when a water sample was apparently empty one day, a few days later it was full of life. Leeuwenhoek had shown that a huge number of minute organisms existed below the threshold of perception, with the theoretical possibility that there might be even smaller 'animals' again, carrying on into the infinitely minute. These incredible observations suggested that life knew no physical boundaries.

One of the things that perplexed Leeuwenhoek was that he could not detect any *young* 'animals' – they all seemed to be the same size, and to increase suddenly in number. This was so unlike normal modes of generation that he said he 'began to think whether they might not in a moment, as 't were, be composed or put together', before concluding, 'But this speculation I leave to others.' Leeuwenhoek later became a fierce opponent of spontaneous generation[35] – and he included in this his tiny 'animalcules' – but his first impression, understandably, was that they were so small and increased in number so rapidly, that perhaps they were assembled instantaneously from the surrounding matter.

The Royal Society was so taken aback by Leeuwenhoek's announcement that when they published his letter in the *Philosophical Transactions*, they took the unprecedented step of inserting a cautionary note: 'This Phaenomenon, and some of the following ones seeming to be very extraordinary, the Author hath been desired to acquaint us with his method of observing, that others may confirm such Observations as these.' In other words, they were not sure it was true. A few months earlier they had been quite happy to print an account of pigs' trotters that glowed in the dark, but they baulked at the idea that there might be incredibly small creatures.[36]

The leading members of the Royal Society not only doubted there really were such minute animals, they also felt that even if they did exist, Leeuwenhoek's estimation of their number must be exaggerated. So they asked the Dutchman for an explanation, and at the same time they instructed Nehemiah Grew to try and replicate Leeuwenhoek's findings on 'pepper-water'.[37] Leeuwenhoek immediately replied, giving calculations to justify his estimate of the number of animals he could see: with around one thousand in a volume the equivalent of a grain of sand, there must be at least one million in a drop of water, he correctly calculated.[38] This was so astonishing it was vertiginous – if there were a million in a drop of water, that meant that in a bucketful, or a pond, there were so many of these tiny creatures that the imagination reeled. Even more hallucinatory than Hooke's close-ups of insects and moulds, Leeuwenhoek's findings showed that the whole planet is teeming with minute life forms.

The Royal Society's consideration of Leeuwenhoek's report was disrupted first by the summer holidays and then by the death of Oldenburg at the beginning of September 1677. By October it was clear that Grew had not made any progress in replicating Leeuwenhoek's observations, and Hooke was asked to take over. In a brilliant insight, Hooke realised that Leeuwenhoek must have put the 'pepper-water' into very fine capillary tubes – 'thin pipes of glass' – which made it much easier to observe the tiny animals. (Leeuwenhoek was particularly cagey about explaining how he did his experiments, and had – deliberately? – neglected to report this decisive point.) At the same time, the Royal Society received a series of testimonials from worthy Dutchmen, confirming that they too had seen Leeuwenhoek's minute animals – as many as 45,000 in a drop of water. After a few weeks of trying various instruments, Hooke was obliged to resort to the single-lens microscope, which he disliked because it was so fiddly. This time the observation succeeded, to the satisfaction of a whole meeting of the Royal Society: 'there was no longer any doubt of Mr Leewenhoeck's discovery,' reads the minute.[39]

Nevertheless, despite this vote of confidence in the man's ability, the Royal Society's credulity was stretched to breaking point about fourteen months later when another letter arrived from Delft, this time written in Latin. Once again it purported to describe incredible numbers of minute organisms that Leeuwenhoek had seen through his tiny microscope. But this time he had not been studying pepper-water, but something infinitely more mysterious: human semen.

In April 1674, Henry Oldenburg had suggested to Leeuwenhoek that he should look at various bodily fluids. Along with blood and sweat, Oldenburg had listed semen as a potentially interesting substance to be investigated using Leeuwenhoek's powerful microscope. With some reluctance (he thought the subject unseemly), Leeuwenhoek had followed his suggestion, but at this stage he had seen nothing more than the 'globules' which he found in all parts of the body. This was not particularly informative, and given that he 'felt averse from making further inquiries, and still more so from writing about them' he did not pursue the subject. Then, in autumn 1677, a young medical student from Leiden, Johann Ham, brought Leeuwenhoek some semen to look at. According to Ham, the sample came from 'a man'. This 'man' had apparently caught gonorrhoea after having 'lain with an unclean woman'. Under the microscope, Ham noticed that the semen was full of 'living animalcules' with tails, which he thought were caused by the disease, through putrefaction. Intrigued, Leeuwenhoek did the obvious thing and used his home-made microscope to look at his own semen.

As he later reassured the Royal Society, this semen 'was not obtained by any sinful contrivance on my part', but was 'the excess which Nature provided me in my conjugal relations'. What this means is that a few seconds before making one of the most surprising discoveries in the history of science, Leeuwenhoek had been making love to his wife, Cornelia. Less than 'six beats of the pulse' after ejaculating, he took some of his semen, scooping it up or squeezing it out, sucked it into a narrow capillary tube, fitted the tube into his microscope, moved to the window or near to a bright candle and pressed his eye close to the minute glass lens. Cornelia's opinion is not recorded.[40]

What Leeuwenhoek saw was to change completely our view of the generation of life: a 'vast number of living animalcules . . . their bodies were rounded, but blunt in front and running to a point behind, and furnished with a long thin tail . . . The animalcules moved forward with a snake-like motion of the tail, as eels do when swimming in water.' These were much larger than the tiny ovals that could be seen in pepper-water, but if anything they seemed to be more numerous – Leeuwenhoek calculated there would be about a million in a volume the equivalent of a coarse grain of sand. That would mean that in an average human ejaculate there would be hundreds of millions of the wriggling things. Most strikingly, they moved with an urgent motion, lashing their tails. They were not like the pepper-water animalcules, gently floating about, moving sedately by fluttering their tiny hairs. Clearly, they were desperate to get somewhere.

Three months after Ham's visit, Leeuwenhoek finally wrote a letter to London. Acutely aware that the topic verged on the obscene, he took two steps in order not to offend. Firstly, he had his letter translated into Latin before sending it; this had the effect of distancing him from the description, ensuring that there was no coarseness of speech and implying there was no impropriety involved. Second, he said to Lord Brouncker that should his Lordship consider it 'either disgusting, or likely to seem offensive to the learned, I earnestly beg that [it] be regarded as private, and either published or suppressed as your Lordship's judgement dictates'.

Brouncker was ill at the time, so the letter was passed to Grew, who had become editor of the *Philosophical Transactions* following Oldenburg's death a few months earlier. There is no sign that Grew thought the subject either disgusting or offensive, but he clearly had his doubts about Leeuwenhoek's report. In a reply written in January 1678, he asked for confirmation of the findings, and suggested Leeuwenhoek look at the semen of animals such as dogs and horses. Leeuwenhoek dutifully went back to his microscope, and then sent two further letters to the Royal Society in March and May 1678, containing more results, which confirmed his original observations.

Curiously, none of these letters (not even the letter of November 1677) was discussed at the Royal Society; the first mention of any of Leeuwenhoek's work on semen occurred nearly twenty months later, on 10 July 1679, and a week after Hooke had introduced Dr Slare, a physician who showed members 'an infinite number of those small wriggling creatures' in horse semen.[41] In fact, the Royal Society seems to have been remarkably unenthusiastic about the whole affair: between October 1678 and October 1679, Leeuwenhoek heard nothing from London, while his first, astonishing, letter was published only at the beginning of 1679. Whether this was due to incompetence, the consistent lack of interest in generation shown by the Royal Society, or the joint disorganising effects of Oldenburg's death and the impact of the arrests linked to the Popish Plot, is hard to say. The fact that Leeuwenhoek continued to send his letters

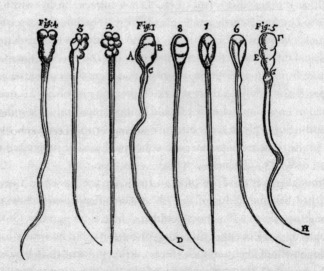

Leeuwenhoek's 1678 drawing of animalcules that he found in semen, as published in the Philosophical Transactions. *Figures 1–4 are from a rabbit, 5–8 are from a dog. Figures 1 and 5 were drawn from life.*

regardless, and did not make contact with the Académie Royale des Sciences in Paris, where Christiaan Huygens was a leading member, speaks volumes for his dogged loyalty to London as his prime channel of scientific communication.

With his second set of observations, completed in February or March 1678, Leeuwenhoek included two drawings executed by one of his Delft draughtsmen. One shows eight of the 'animalcules' seen in the semen of a dog and a rabbit. To the modern eye these are clearly spermatozoa; to Leeuwenhoek, they were simply tiny animals, no different from the incredibly small life forms he had found in pepper-water. The object in the other drawing, which he mistakenly thought was his most important discovery, is more mysterious. In his letter to Lord Brouncker, Leeuwenhoek had said that the 'denser substance of the semen' consisted mainly of a tangle of tiny vessels which he took to be arteries, veins and nerves and which were so densely packed that he wrote: 'I have seen so many of these vessels that I believe I have observed more in a single drop of semen than an anatomist encounters in a whole day of dissection.' The accompanying drawing duly shows a complicated crisscross mess of tubes.

Leeuwenhoek was so taken by the apparent complexity he thought he could see in these 'vessels' that he declared, without any further evidence, that 'it is exclusively the male semen that forms the foetus and that all the woman may contribute only serves to receive the semen and feed it'.[42] This was doubly shocking. Firstly, Leeuwenhoek gave no reason to abandon the now widely held idea that eggs played a decisive role in generation. This was not 'proof' as science was coming to expect. It was an idea, a hypothesis, a theory; but it was not a fact. Second, it was not at all clear that the 'vessels' which inspired Leeuwenhoek's breezy confidence actually existed. Grew, in particular, was not impressed and suggested that they might have been produced by some kind of evaporation. At first Leeuwenhoek vigorously rejected this suggestion, but by 1683 he accepted that Grew's doubts were justified and that this particular observation was 'merely accidental'.[43] Meanwhile, these 'accidental' vessels – the exact nature of which is still unclear – distracted his attention from the truly important part of his discovery.

Long before these exchanges between Delft and London were published at the beginning of 1679, news of Leeuwenhoek and Ham's discovery began to circulate in Europe, and people tried to see if they, too, could observe the tiny animals. They could. On 20 January 1678, in an underlined postscript in a letter to Thévenot, Swammerdam wrote: 'In the semen of a mouse and a fairly big dog we have observed innumerable small worms, shortly after the excretion'.[44] A few weeks later, Swammerdam dissected a cuttlefish and found a great many 'delicate white parts' in its testicles which, to his amazement, began to move when he put them in water. His detailed drawing, made at the time but not published until 1737, shows a cuttlefish 'spermatophore', which he noticed released 'a little surprising moving white particle'; this was presumably a spermatozoon.[45] At about the same time, the young Danish anatomist Thomas Bartholin examined human semen under the microscope and reported seeing many tiny animals in it, while at the end of August 1678 the Dutch physician Nicolas Hartsoeker sent a letter to Paris in which he reported that in the semen of a cockerel he had observed creatures like 'new-born frogs', which could also be found in the semen of other animals.[46]

The first published description of Leeuwenhoek's findings was made (without naming their author) by Christiaan Huygens, who had received a copy of Leeuwenhoek's original letter to Brouncker. After demonstrating the existence of the animals in semen to a meeting of the Académie Royale des Sciences in summer 1678, Huygens made a summary of Leeuwenhoek's methods and findings, which was inserted in the last two pages of the *Journal des Sçavans*. Most of the article deals with the single-lens microscope and Leeuwenhoek's discovery of animalcules in pepper-water, which was arresting enough; but the last few sentences hint that Huygens may have grasped the potential significance of Leeuwenhoek's observations of semen:

> It could be objected that these animals are generated by corruption or fermentation: but there is another sort that must have another origin.

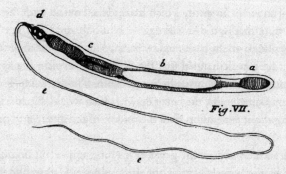

Swammerdam's 1678 drawing of a cuttlefish spermatophore, published in 1737. Swammerdam apparently observed sperm in the region marked b.

Such as those that can be found with this microscope in the semen of animals, which seem to be born with the semen, and which are present in such a large quantity that it seems it is virtually entirely composed of them. They are all made of transparent matter. They have a very rapid movement, and they look similar to frogs before their legs have formed. This latest discovery, which has been made in Holland for the first time, seems to me to be extremely important and will provide material for those who seriously study the generation of animals.[47]

The far-reaching scientific implications of finding animalcules in semen appear at first not to have occurred to Leeuwenhoek – he made no mention of this in his original letter, apparently assuming that they were merely another example of the minute life forms that could be found everywhere. His main preoccupations appear to have been the non-existent 'vessels' and the potential embarrassment that might be caused by studying such an intimate matter. Like Huygens, the leading members of the Royal Society were rather sharper. More than six months earlier, in his January 1678 reply to Leeuwenhoek's first letter, Grew had immediately assumed that these 'animals' were directly related to generation. He had asked Leeuwenhoek how he thought his discovery fitted with the widely accepted model according to which semen was merely 'the vehicle of a certain highly ethereal and volatile spirit, which makes vital contact with the site of conception, that is, the female egg'.[48] Not only did Grew imply that Leeuwenhoek might have made a decisive discovery relating to generation, he also demanded that the Dutchman explain the apparent contradiction of the prevailing egg-centred understanding of the question. Battle lines were being drawn which would divide scientific thinking about generation until the middle of the nineteenth century.

Having initially missed the point of his most important finding, distracted by the 'vessels' which only he could see, Leeuwenhoek soon focused his attention on the meaning of the tiny 'animals' he had found in sperm, and on their role in generation. Any lingering thought that they might be produced by decay or disease was soon banished by the fact that he found them in every kind of semen he studied, from codfish to cat, from flea to frog – he even speculated that the creatures he found in pepper-water might reproduce through their own, even smaller, seminal animalcules.[49] Dissections of dogs and hares proved that, far from being generated in the penis, as he originally thought, the little animals could be found virtually everywhere in the testicles.[50] These strands of proof all contributed to his growing conviction, first expressed in January 1683, that 'man comes not from an egg but from an animalcule that is found in male sperm'.[51]

Leeuwenhoek was well aware that most people did not agree with him. Grew had not only asked for a general comment about the mode of action of semen, he had also pointedly asked Leeuwenhoek to explain the key finding that had shaped the commonly held understanding of generation: the absence of semen in the uterus, as reported by both Harvey and De Graaf. This well-supported observation apparently limited semen to acting on the egg indirectly; in the dominant Harvey/De Graaf vision of generation there was no place for little animals in the semen. Leeuwenhoek initially put forward two explanations: perhaps the female deer and rabbits studied by Harvey and De Graaf had voided all the semen out of fear when being killed, or perhaps it had evaporated when the animals were dissected. While both of these interpretations were potentially true, they had not been demonstrated and were therefore unsatisfactory. However, Leeuwenhoek soon came up with a far more convincing answer which was surprisingly straightforward: both Harvey and De Graaf were wrong.

Leeuwenhoek made his discovery in the simplest way imaginable. Having abandoned his ideas about the 'vessels', he decided to see how long the 'little animals' could stay alive. To his surprise, he discovered that if they were kept warm, those in dog semen could survive for up to seven days. He then took a bitch on heat, allowed her to mate three times over the course of a day, and had her killed. Although there was no sign of semen inside her when he looked with the naked eye, under the microscope he found thousands of his eel-like creatures in the uterus, while he found no sign of anything like an egg. Hampered by their lack of an appropriate instrument, Harvey and De Graaf had missed the truth: after mating, what now appeared to be the decisive component of semen – the little animals – could be found in the 'horns' of the uterus, near the openings of the Fallopian tubes.

This led to Leeuwenhoek's developed vision of generation in humans and other animals with internal fertilisation: the seminal animals would swim into the uterus, up into the 'horns', where they would find a 'few little veins' on to which one or occasionally more of the little creatures would attach themselves and feed.[52] They would then grow, shedding their tails (supposedly like tadpoles lose their tails as they turn into frogs), and would each 'coagulate into a round ball or globule'.[53] The structure which Harvey, Swammerdam and De Graaf had found in the uterus and had taken to be an egg, was in fact 'a body animated by a living soul, derived from the male seed'.[54]

Leeuwenhoek's theory not only marked a radical change from the recently accepted understanding of the role of the ovaries and their vesicles, it also redivided the living world along the lines of oviparous and viviparous organisms. Up until his discovery, the work of the generation network had consistently shown that all animals – from lice to lizards and from mice to men – came from eggs. Leeuwenhoek's vision of generation suggested that egg-laying animals were indeed fundamentally different from viviparous organisms, because they had completely different ways of feeding the spermatic 'animalcule', which was the sole cause of generation. The obvious eggs of chickens, lizards, flies and the like were the primary food source of the 'little animal', whereas in humans and other mammals the animalcule was somehow nourished in the uterus.

To find real evidence in favour of his theory of generation, Leeuwenhoek began a series of anatomical and microscopical studies of ovaries and their function. Having apparently seen both the dissections and the drawings of human ovaries made by Swammerdam and De Graaf at the beginning of the 1670s, he was surprised when he looked at ovaries in humans, cows and sheep.[55] The vesicles did not hang free like grapes, as Swammerdam's illustration implied, but were solidly embedded in the flesh of the ovary. The suggestion that the vesicles could somehow 'get out' of the ovary, he argued, was 'altogether at variance with nature'. When he did observe ruptured follicles, he dismissed them as examples of 'ulceration',[56] concluding scornfully that 'these pretended eggs were nothing but the discharge of some vessels'.[57]

The obvious solution to this dilemma was proposed in October 1679, when Hooke suggested that Leeuwenhoek should look inside fertilised and unfertilised mammalian 'eggs' (in fact, ruptured and ripe follicles, respectively) and see if he could detect a difference. Leeuwenhoek tried this on a number of occasions in the late 1670s and early 1680s, but he found nothing striking; all he could ever see was a 'thin, watery substance, mixed with some globules'.[58] Even when he tried to replicate De Graaf's study of generation in rabbits, he still found no sign of the egg. This is surprising, because the true egg (the oocyte, which is released when the ovarian follicle bursts) would have been easily observable with his microscope; it is about a hundred times larger than a single sperm. Perhaps it was one of the 'globules' Leeuwenhoek saw, or more likely it simply trickled out unnoticed when he punctured the follicle. Whatever the case, he did not see what was literally under his nose.

It is tempting to view this as one of science's great lost opportunities, as a 'what if' moment that could have changed the course of human knowledge, or at least short-circuited the 150 years of confusion which followed before a full scientific understanding of conception was developed. But as much as anything can be certain when it comes to this kind of speculation, such a conclusion would be completely mistaken. Leeuwenhoek knew that eggs were involved in generation in birds, insects and other oviparous animals, and he was brilliant enough to suggest that there was a site on the egg where a single sperm enters, which he discussed with regard to fish, frogs and fleas. But the profoundly mistaken conclusion he drew was that the egg was merely food. He was so convinced by his idea that the 'little animals' were the basis for the future embryo that he minimised the role of the egg even in egg-laying species. Everything was seen through the distorting lens of his semen theory of generation. Had he seen the mammalian egg, there is little reason to imagine he would have thought it any different from a chicken egg: he would have seen it as nothing more than nourishment for a lucky little animal.

Leeuwenhoek's final view of the egg theory of generation, expressed in the mid-1680s, was extremely scornful: he described it to the Royal Society as 'one of the most addle-pated propositions current among physicians'.[59] Indeed, he found the whole idea so ridiculous that he said that were Swammerdam and De Graaf still alive they 'would blush with shame about their imaginings'.[60] Leeuwenhoek's open hostility to the egg theory flowed from a number of factors. Apart from seeking to defend the importance of his discovery of the tiny creatures in semen, he felt there was a fundamental difference between the spermatic 'animals' and the egg. Eggs were immobile and showed no sign of life, whereas the little animals in semen wriggled and squirmed, clearly demonstrating – to him at least – that they were alive and had a soul. This must mean that the animal, and not the egg, was the key factor in generation. As he put it in 1685: 'Surely no one will question that an animate creature, however small it may be in the womb where it is nourished, is endowed with a living soul. This being so, it is surely a thousand times more probable that the living soul, possessed by the animalcule of the male seed, remains in it, and that the animalcule only changes its shape when the living soul passes from the male seed into another body.'[61] At a time when movement was equated with life, this was a powerful argument.

Above all, Leeuwenhoek raised a series of criticisms of the evidence traditionally presented in favour of the egg theory. These ranged from minor quibbles such as the presence of vesicles in the ovaries of immature female mammals, and the very real problem of finding nothing in the 'eggs' except fluid and a few globules, to the fact that there was no solid explanation of how the eggs got from the ovary into the Fallopian tubes. This was widely accepted as being the biggest problem with the egg hypothesis – both Swammerdam and De Graaf had recognised as much.

Leeuwenhoek approached the question in typically robust fashion, unfairly suggesting that the supporters of the egg hypothesis thought that what he called the 'eggs' (in fact the ovarian vesicles) were 'sucked' or 'pulled' off the surface of the ovary by the end of the Fallopian tube.[62] This description, which appears to have been Leeuwenhoek's own, was both a reasonable and a one-sided way of describing contemporary opinion. It was reasonable because it was as good a hypothesis as any (no one had any clear idea of what actually happened); it was one-sided because neither De Graaf nor Swammerdam ever suggested anything of the sort – their explanations were much more tentative, involving movements of the Fallopian tubes during copulation or changes in the position of the organs after death. The idea that the floppy Fallopian tubes could generate the force required to remove a vesicle was simply ludicrous, as Leeuwenhoek must have recognised. But although this argument was clearly a straw man, the people who thought the vesicle was the egg had no clear and agreed alternative hypothesis.

Even if Leeuwenhoek was convinced that viviparous organisms did not produce eggs, he still had to explain the function of the ovaries. He vaguely suggested they were 'merely instruments serving to relieve certain adjacent parts',[63] before finally concluding, with Aristotle and Harvey, that they had no function; like male nipples, they were the functionless equivalent of structures that were vital in the other sex.[64] Some of his supporters, however, were less convinced, pointing to evidence that suggested that the ovaries and their contents, the eggs, at least had a nutritional role in viviparous animals, and in particular in humans.

In 1693, the Reverend George Garden of Aberdeen sent a letter to Leeuwenhoek outlining why he favoured the idea that the animals in human semen were nourished by a product of the woman's ovary. Garden sought to re-establish the link between viviparous and oviparous animals which had been broken by Leeuwenhoek, arguing mistakenly that in humans, as in chickens, there was a zone on the surface of the egg which would allow only one little spermatic animal to enter, in order to feed. The only difference between viviparous and oviparous animals, argued Garden, was the size of the eggs, which, being smaller in viviparous organisms, meant 'that the Foetus must send forth roots into the Uterus to get its nutriment'.[65] Imagining that the 'animalcule' had to meet the egg in or next to the ovaries (fertilisation in fact takes place in the Fallopian tubes), he also dealt with the thorny problem of the distance between the ovaries and the entrance to the Fallopian tubes by suggesting that the two get closer during copulation.

Apart from these reinterpretations of accepted facts, Garden challenged Leeuwenhoek with two important arguments. Firstly, a strictly spermatic theory of generation such as that advocated by Leeuwenhoek could not explain why, if the ovaries were removed from a female animal, it could no longer conceive. Second, Garden thought it very unlikely that, as Leeuwenhoek argued, an animalcule could metamorphose into an egg, like a tadpole into a frog, without feeding – in the initial stages of embryonic growth there was no sign of any kind of attachment to the uterine wall.

Leeuwenhoek's reply was marked by his usual mixture of confidence and breezy dismissal of apparently contradictory findings. He simply ignored Garden's reminder that De Graaf had shown that 'those thin membranes in the Ovary are not the Eggs, but merely serve to form the glands in which the Eggs are formed, which break because a thin papilla in the gland opens'[66] – in other words, the real egg was in the fluid inside the vesicle, not the vesicle itself – thus making it a bit easier to imagine how the egg might get down the Fallopian tube. Instead, Leeuwenhoek reasserted that the 'imagined Eggs' were 'very firmly enclosed in a membrane', and claimed that when he looked at bitches or rabbits on heat he was unable to see 'in any place in the Ovary a sign that an imagined Egg might have been removed'.[67] De Graaf's detailed experiments were not mentioned.

In answer to Garden's suggestion that the animalcules required an egg to grow, Leeuwenhoek replied that 'many among the animalcules in the male seed which are poured into the Uteri' might in fact start developing into embryos, without being detected because they were simply too small – even if each animalcule grew a thousandfold, it 'could not yet attain to one hundredth of the size of one grain of coarse sand'.[68] He then questioned whether it would be possible to find such a tiny structure 'among all those viscous and coagulated particles with which the Uteri are found to be covered on the inside'. Nine years earlier he had had no such doubts about the resolving power of his microscope, when he described his dissection of a bitch on heat shortly after mating and stated with absolute certainty that there was nothing in the 'horns' of the uterus that resembled an egg – 'Had there been one particle in them no bigger in size than the hundredth of a grain of sand,' he declared, 'I doubt not but I should have found it.'[69] Leeuwenhoek could be strikingly inconsistent in his arguments.

Leeuwenhoek's explanation of the sterility of animals that had had their ovaries removed was equally unsatisfactory, avoiding the argument. During the removal of the ovaries, he claimed, 'many vessels are injured which were destined not only to supply the nutriment for the growth of the Animals lying in the Uterus, but also to sustain and nourish the Uterus itself'.[70] However, he backed this up with a shrewd observation: the uterus shrinks in animals without ovaries, which could explain why they could not conceive. In fact this is due to hormonal changes produced by removal of the ovaries, but at the time Leeuwenhoek's explanation was entirely consistent, and effectively deprived Garden of one of his more powerful arguments.

Nevertheless, most thinkers remained unconvinced by the sperm theory of generation. Although the Amsterdam clergyman Benedictus Haan supported Leeuwenhoek's ideas (he had been one of the witnesses who had written to the Royal Society over the animalcules in pepper-water), and said he found it astonishing 'that there are still People who uphold the insane opinion about the Female Ovary',[71] he was very much in a minority. Up until the end of the seventeenth century, the prevailing view was expressed by John Ray, who baulked at the incredible numbers of spermatic animals that were apparently wasted in each episode of generation. In a provocative and light-hearted calculation, Leeuwenhoek had worked out that there were about ten times as many 'little animals' in the ejaculate of the average codfish than there were humans on the face of the planet.[72] (Given that he substantially overestimated the seventeenth-century global population of humans as 13.4 billion – the world population is still under half that figure – the truth was even more staggering than he thought.) If each of these animalcules could potentially become a living individual, that posed a number of philosophical and theological problems. In 1693 Ray explained that he disagreed with Leeuwenhoek's ideas about the role of the seminal animals in generation 'because of the necessary loss of a multitude, I might say infinity, of them, which seems not agreeable to the Wisdom and Providence of Nature'.[73] Leeuwenhoek's response to this criticism, which he had encountered long before Ray came up with it, was to point to the incredible number of seeds that were produced by plants, but which nonetheless gave rise to only a few seedlings.

Nevertheless, while the plant analogy carried some weight, the problem remained. As Leeuwenhoek himself had pointed out, the wriggling animals in semen were different from eggs, or seeds – they were clearly alive. The idea that all those actual living things might be wasted was shocking to the theologically inspired thinking of the time. The consequences were even graver if the full implications of Leeuwenhoek's arguments were applied to the animalcules in human semen: if each animalcule was alive and was a potential human being, that meant it had a soul. As the French physician Claude Brunet put it in 1698, Leeuwenhoek's doctrine accused 'the sovereign Ruler of having carried out an infinite number of murders or created an infinite number of useless things by forming in miniature an infinite number of men destined never to see the light of day'.[74] In the Dutch Republic, with its traditions of religious toleration, Leeuwenhoek's theologically dangerous ideas did not get him into trouble. Had he been working in Italy, things might not have been so straightforward.

Leeuwenhoek's investigation of the microscopic world ceased only when he died, aged ninety, in August 1723, fifty years after he had been introduced to the Royal Society by De Graaf. His daughter Maria sent a cabinet full of his microscopes and samples to London where, as with Swammerdam's preserved uterus, they eventually went missing.[75] A few examples of his delicate instruments survive in museums, although the viewer naturally concentrates on the least important part – the silver plate and the screw-based object holder. The truly vital component – Leeuwenhoek's minute glass-bead lens – is often overlooked. But it was this small piece of glass that enabled him to see so far, and to produce observations that were not surpassed for well over a century.

In a way, Leeuwenhoek's findings, and above all his interpretations, struck a chord because they gave new vigour to old ideas. From the very beginnings of human thought about generation, male semen had played a decisive, even magical role. For Aristotle it had been the key feature of generation, the active factor that shaped the female blood. Semen had also been at the heart of Paracelsus's recipe for generating a human being, while Harvey, like Aristotle, had argued that semen is 'imbued with formative power, that is to say is spirituous, operative and akin to the element of the stars'.[76] Leeuwenhoek's findings suggested that if such a mystical power existed in semen, it was localised in the animalcules. Furthermore, with his theory of generation based on the sole power of sperm, he unintentionally echoed the old Aristotelian position of a single 'semen', with the female merely providing the support and nutrition required for its growth.

There was also a less direct link with old ideas. Alchemists assumed there was a parallel between the male ejaculate and the apparent ability of all of nature to produce change – both were called 'semen' or 'sperm'. The alchemists thought that even inorganic substances like metals, which could become liquid when heated, could form alloys, or could show astonishing and unexpected interactions, also had a 'sperm'. In the early years of the seventeenth century this concept was used by the new wave of 'atomists' such as the French thinker Pierre Gassendi, to describe the action and power of the 'molecules' they thought made up the universe. The transformation of alchemy into chemistry in part relied upon the old pre-scientific idea of 'sperm'.[77]

By the second half of the century, faced with the growing power of modern scientific knowledge, alchemy was becoming less and less popular, although figures such as Boyle and Newton continued to try and interpret aspects of the natural world in the old way. Even as late as 1672, the King's physician, Thomas Sherley, felt able to proclaim that 'As Vegetables, and Animals have their Original from an invisible Seminal Spirit, or breath; so also have Minerals, Metals, and Stones.'[78] For the alchemists, the whole of the Universe was pulsating with a life force. Leeuwenhoek, they could argue, had merely revealed some of the tiny 'seminal particles' which supposedly pervaded the Universe. These particles had been at the heart of the Royal Society's early discussions of spontaneous generation in the 1660s, and were particularly dear to Robert Boyle. Although these parallels between the latest scientific discoveries and alchemy found little explicit expression at the time, they show how the development of science in the seventeenth century did not simply break with the past, but reworked and remoulded old ideas in the light of new facts and new hypotheses.

Using his microscopes, Leeuwenhoek had found the answer to the problem that had so perplexed Harvey and De Graaf: semen could play a direct, physical role in generation, through the action of the tiny animals wriggling their way into the uterus. He had blown apart the egg-centred consensus about generation created by the work of Steno, Van Horne, Redi, Swammerdam, De Graaf and Kerckring. They had looked at the role of the female; Leeuwenhoek had focused on the other side of the generation equation. Following these contrasting approaches, throughout the eighteenth century biology was divided into two camps – the ovists and the spermists, the names reflecting the structure each group thought was fundamental in generation. The issue became one of the great battlegrounds in science, attracting the attention of philosophers and writers as they sought to grapple with the implications of each side's inadequate evidence.

During the initial development of the two views of generation, neither side actually provided much in the way of direct proof. De Graaf and Swammerdam had produced some accurate observations and, in the case of De Graaf's work on rabbits, some telling correlations between the number of burst follicles and the number of embryos in the uterus. But they had not actually demonstrated that the egg (whatever exactly that was) turned into the embryo, nor had they done any experiments of the kind that had made Redi's work so influential. Leeuwenhoek's position was even weaker: he had taken a stance against the prevalent egg theory of generation for a series of reasons that most observers found partial and unconvincing. His certainty was based above all on a mixture of brash self-confidence, real problems with the egg theory and the enigma of the squirming seminal animalcules. But his arguments generally involved *post hoc* justifications of his positions, few verified predictions and fewer experiments. Under these circumstances it is no wonder that neither side was able to land a killer blow. There was simply not the level of proof necessary to convince everyone that one side or the other was right.

FROM GENERATION TO GENETICS

The torrent of work on generation that took place between 1665 and 1680 produced what looks like a perplexing paradox. Two visions of generation appeared where, to modern eyes, there should have been only one. Instead of realising that egg and sperm were the two complementary components – each representing half of the information required to make a new organism – thinkers argued that although both elements were necessary, one of them (either egg or sperm) was the key factor. For the ovists, the semen merely awoke the egg; for the spermists, the egg (in those species in which there clearly were eggs) was simply nourishment.

The 'two-semen' model of generation advocated by Galen could have provided a correct interpretation of the discovery of egg and sperm. But the very discoveries that Galen's idea could have explained led to that idea being abandoned. Galen had argued that males and females made equivalent contributions to generation, but De Graaf and Swammerdam had shown that female animals, and in particular women, did not produce anything like male 'semen', but instead something they thought was more like a tiny bird's egg – but without the shell. Far from being equivalents, the two substances that had been discovered appeared to be completely different in every respect. For example, in chickens the egg was single, large and apparently passive, while the semen contained an astonishing number of tiny, very active spermatic animalcules. From this point of view it is hardly surprising that no one argued that these were complementary components.

For nearly 180 years the debate between ovists and spermists divided thinkers all over the world. Although significant findings relating to generation were made in the eighteenth and early nineteenth centuries, they did not immediately give rise to a new, correct understanding of the respective roles of egg and sperm. Substantial scientific progress in understanding generation did not occur for many years – at least until the 1850s, and, at a more fundamental level, until the discovery of the genetic code which occurred a century after that.

Looking back at the outcome of the work of the generation network in the 1660s and 1670s, and comparing this spasm of productivity with what appears to be the long impasse that followed, it might seem that the thinkers of the time were stupid or ignorant, refusing to see what was obvious – that egg and sperm both contribute to the creation of a new organism. That would underestimate both the many intelligent people who held these views over such a long period, and the brilliance of the scientists who eventually brought humanity to our current state of knowledge, who transformed our fragmentary understanding of generation into the modern sciences of genetics, reproductive biology and development. It would also mean that we do not need to try and understand what framed and limited the thinking of previous centuries. But, of course, those early scientists were not stupid. Our challenge is to understand how they could at the same time be so clever and so wrong. Resolving this enigma can also provide some insight into what shapes our current scientific ideas and theories, many of which will turn out to be at least inadequate, and some of which may be completely mistaken.

*

The main reason why we think that seventeenth-century scientists ought to have understood the roles of egg and sperm is that we look at the problem through modern eyes. The gulf between our present understanding of the natural world and that which existed in the seventeenth century is so deep that it requires a huge effort to build a mental bridge back to those times. We know that in 'generation', genes – the units of inheritance – are transmitted in egg and sperm, and together provide the necessary information to make the new organism, in conjunction with appropriate environmental conditions. Because of the central role of genes in this process, we realise that you cannot understand generation without understanding heredity. But this focus on heredity simply did not exist in the seventeenth century, nor could it, because it was not clear that anything like 'heredity' – a consistent pattern of relations between parent and offspring – actually existed. As a result, there was no word for it. Although the word 'hereditary' was in use at that time (for example, Bacon wrote that 'Long Life, is like some Diseases, a Thing Hereditarie, within certaine Bounds'[1]), 'heredity' was not used in its biological meaning until the 1860s.

Hard facts about the similarities between parents and offspring were difficult to come by, and even more difficult to interpret and develop into a general theory. A child might look like its mother, but prior to the work of the generation network even learned people accepted that in some cases a woman could give birth to a rabbit or a kitten. Most people who studied generation therefore avoided dealing in any detail with the relations between parent and offspring, and nobody carried out a systematic study of the phenomenon. When they did broach the subject, their conclusions revealed the depth of the confusion that existed at the time.

De Graaf provided some telling examples of hereditary phenomena which underlined how difficult it was to interpret what was going on: 'If a mare receives the semen of an ass,' he wrote, 'the foetus has the features not of the father alone but those of both parents mixed together; likewise if a goat penetrates a sheep, a sheep with rather coarse wool is produced. Similarly, the offspring of a white man and an Ethiopian woman acquire a mixed colour.'[2] In other words, characters appear to be a blend of paternal and maternal contributions. As De Graaf pointed out, the 'two semens' tradition of Galen and Hippocrates had no difficulty in explaining this: the foetus 'is as much like its mother as it is its father; mothers, therefore, as well as fathers have child-producing semen'. But as De Graaf recognised, the Aristotelian tradition, which argued that the male semen determined the form taken by the female menstrual blood, was equally able to explain how the female's contribution could sometimes have an effect, and thus account for apparently mixed inheritance. Aristotle himself had dealt with this issue, pointing out that plants grown from the same seed but on different soil could look very different. Although this did not really explain what was going on, the fact that Aristotle had an answer was enough to satisfy De Graaf, who felt that both models were able to account for these observations.

In fact, none of these phenomena were quite so simple as they might appear. Even the apparently straightforward question of the relation between the skin colour of parents and that of offspring appeared complex 350 years ago. In *Miraculum Naturae*, Swammerdam gave another explanation of how skin colour could be determined, based on the case of a Dutch child with dark patches of skin. This example was so striking that it was highlighted in the *Philosophical Transactions*: 'a woman at Utrecht . . ., being surprised with the sight of a Negro, and so exceedingly frightened as to become speechless for the time; had a strong fancy she should bring forth a black child'. This was not simply a piece of ignorant nonsense – the idea that the experience or imagination of the pregnant woman could profoundly affect the offspring, which went back to Galen, was widespread in the seventeenth century. The Dutch Golden Age diplomat and poet Jacob Cats wrote:

> When a woman is pregnant, the man should take care
> That nothing is found in his house

That is deformed, a gruesome or strange sight,
To aggravate our eyes and so distract our senses.[3]

According to Swammerdam, the woman decided that by washing herself with hot water she could 'wash away the blackness' of her unborn child. 'She was at length delivered of a child that was indeed white, yet those parts excepted, where the water in the washing had not touched; such as the interstices of the fingers and toes and some other places, where the manifest tokens of blackness appear'd.'[4] Swammerdam's example suggested that at least some characters might be affected by environmental conditions, which made it even more difficult to imagine there could be a coherent theory of heredity.

Furthermore, despite De Graaf's instances of 'blending' inheritance, it was quite apparent that not all characters followed this pattern. Children sometimes looked like one parent, sometimes a mixture of the two, and sometimes like neither. Sometimes characters seemed to skip a generation, and children looked like their grandparents. In *De Generatione*, Harvey highlighted all these examples, wondering why 'the foetus is sometimes more like the father and sometimes the mother, sometimes the grandparents either on the father's or the mother's side.[5] Shrugging his shoulders, Harvey went on to deal with some other equally perplexing examples, such as the fact that birthmarks and moles can sometimes be inherited, and sometimes not. Finally, he underlined the striking contrast between the inheritance of most characters, which generally involves some kind of blending of the mother's and father's make-up, and the inheritance of sex, which in virtually all cases shows no blending at all: 'Why is it that though we see the offspring has for the most part a mixed make-up or composition derived from both parents in all the rest of its parts, yet it is not so in the genitals, for most are born either male or female, and very seldom hermaphrodite?'[6] This was an extremely telling comment, for it showed that a decisive element in the make-up of any individual – its sex – was determined in a way that was qualitatively different from that of most other characters.[7]

As if the example of sex was not enough to confuse people, thirty years later Leeuwenhoek discovered another problem – the first instance of what geneticists call a 'dominant' character. Leeuwenhoek pointed out that all wild rabbits were grey, but that if a wild male was 'mated not only with white, but also with piebald, bleu and black does . . . all the young issuing from this, take their father's grey colour'. 'Indeed,' he went on, 'it has never been seen that any such young rabbit had a single white hair or any other hair than grey'.[8] His ingenious explanation of this phenomenon was completely logical, and completely wrong: he argued it was 'a proof enabling me to maintain that the foetus proceeds only from the male semen and that the female only serves to feed and develop it'.[9] In other words, there was no relation between both parents and the offspring, but simply between father and offspring, represented by the little animal in the male semen. The father was grey, so the offspring was inevitably grey, thought Leeuwenhoek. (Had he mated a grey female with a white male, he would still have got grey rabbits, which might have made him think again.)

Because of the confusing impact of these contradictory examples and explanations, no one realised the implication of the discovery of the two component parts of what we call reproduction. The relations between parent and offspring appeared contradictory, and could be more or less adequately explained by theories based on only one decisive component. Neither the facts of generation, nor their theoretical interpretation, provided a compelling reason for people even to realise that heredity existed, far less to develop a new understanding of it.

A related reason which made it difficult to think along the right lines was society's overall level of technical development. Scientific explanations of the natural world generally use technological analogies. The very act of discovery implies that something new has been observed, for which there is no immediate explanatory framework. To describe something for which there are as yet no words, scientists are obliged to find analogies from everyday life. The use of analogies extends even further than the description of novel objects and phenomena – they frame the whole outlook of science. For example, 350 years ago, mechanics were at the cutting edge of human technology. By harnessing various sources of power – mainly springs, water or wind – machines could be made to measure time, to process food or to pump water. The mechanisms involved in such machines were relatively easy to understand, even if the nature of the forces that powered them (pressure, kinetic potential etc.) remained obscure. The analogy of clockwork was used by astronomers to explain the movement of the stars – they even built celestial clockwork models, or orreries, to model the orbits of the planets. Natural historians also looked to the most advanced contemporary technology for inspiration, and viewed living organisms as divinely created machines, powered by the kinds of forces that were involved in manmade mechanisms. This approach was extremely productive when it came to explaining some things – for example, Harvey's analogy of the heart as a pump, or Steno's 1667 study which involved a mechanical interpretation of muscle function.

Paradoxically, these successes also underline the limits of the mechanical analogy. Harvey was able to portray the heart as a pump because that is what a heart is; similarly, muscles function by exerting mechanical force on bones, and their behaviour can be explained using the laws of mechanics. But when it came to heredity, purely mechanical analogies failed because heredity does not behave according to simple mechanical principles. A single mechanical analogy cannot explain the very different examples of blending inheritance, dominance and sex determination.

On a more general level, the success of mechanical analogies actually hindered appreciation of the complementary roles of the male and female components of conception. Because it was thought that organisms were like clocks, a correct understanding of generation would require people to imagine two clocks splitting in half, and then each pair of half-clocks somehow coming together to make a new clock. This would have required not only an unimaginably complex process, it would also have raised the question of how the two half-clocks 'knew' which bit went where. In the seventeenth century this would inevitably have meant a return to Aristotle's 'final cause' explanation. The search for a mechanical analogy would have led right back to a supernatural mind-set – exactly what the scientific revolution was trying to replace.

Understanding the process of generation as we now do would have been impossible 350 years ago. The very concepts we use to explain inheritance and genetics – transmission, information, programme, code – are the product of the electronic age and were consistently applied to understanding generation only after their widespread adoption in the early years of computing around the Second World War. Although a seventeenth-century scientist would have understood what a code was, the idea that egg and sperm each contained half the necessary information to make a new organism, and that this genetic code was the essential thing transmitted from generation to generation, would have been so much technobabble.

There is an intriguing corollary to all this: computers are currently the most advanced form of our ability to manipulate matter, and concepts such as information, programming and feedback loops are an integral part of modern attempts to model and explain biological phenomena. Today it is impossible to imagine anything richer and more powerful than this model. But eventually, following new and unforeseeable technological developments, this approach will no doubt seem quaint and naïve. The future will prove that we have a vision that is limited by the boundaries of our scientific imagination, which in turn is largely determined by our social conditions, by the way production is organised, and in particular by our technology.

The final element which explains why science split into 'ovists' and 'spermists' can be found in an aspect of generation which was not widely studied in the seventeenth century, but which was to be of decisive influence over the next 150 years. The work of Swammerdam, De Graaf, Redi, Leeuwenhoek and the others had shown that all animals come from the egg of an animal laid by the same species, that like breeds like, and that both male and female produce special structures which they associated with generation. In this they were truly revolutionary. But when it came to the question of what there actually was in an egg (or a sperm), they merely extended existing thinking rather than challenging it. This issue – the origin of organic structures – reinforced the ovist–spermist fault line that ran through the whole of scientific thinking about life, and which, in a certain manner, still exists today.

The traditional, pre-scientific view of generation emphasised the role of the male 'seed', and concluded that it must in some way contain the future organism. According to the Roman writer Seneca, 'In the semen there is contained the entire record of the man to be, and the not-yet-born infant has the laws governing a beard and grey hair. The features of the entire body and its successive phases are there, in a tiny and hidden form.'[10] In a similar vein, more than 1500 years later, the alchemist Robert Fludd (1574–1637) wrote: 'in the seed or Sperm, though nothing do appear in the first degree explicitly but Sperm, yet the whole man, namely the bones, flesh, blood, sinews, and such like, are complicitly conteined, and will by degrees appear out explicitly'.[11]

In one way or another, all the seventeenth-century scientists who made such progress in the study of generation agreed with this view, which was subsequently known as 'preformation' and which implied that in some way or another the future organism lay preformed in either the egg or the sperm.[12] But although this was everyone's underlying assumption, most of those who were interested in generation usually avoided the question (neither De Graaf, Steno nor Redi, for example, ever wrote about the subject) because, given the tools available, trying to find out what there was in an egg (or a sperm) inevitably involved speculation rather than detailed observations. The scientists of the time were consciously trying to root their theories and descriptions in facts; the question of precisely where the fully grown organism came from was not worth spending a great deal of time on, for no matter how closely you looked at an egg (or a spermatic animalcule) it was impossible to see anything with any certainty.

Not everyone agreed, however. Nicolas Hartsoeker, for example, knew exactly how a human being was generated from sperm. Hartsoeker was a bit player in the generation network who claimed (falsely) to have observed spermatic animalcules long before Leeuwenhoek.[13] In 1694 he published a drawing that has been repeatedly reproduced down the centuries,[14] and is still sometimes used to indicate how foolish seventeenth-century scientists could be; in fact, it simply shows how ignorant subsequent critics have been. Hartsoeker's famous figure shows a tiny man, or homunculus (although neither he nor anyone else at the time used this loaded term), fully formed and curled up in the head of a spermatozoon. Ingeniously, Hartsoeker suggested that the little animalcule connected with its food tail-first, with the tail going on to form the umbilical cord. The overall effect is a drawing that looks more like an alchemical or mystical portrayal of generation than anything that can be seen down a microscope, and for much of the twentieth century students of embryology were encouraged to snigger at it.[15]

In fact, Hartsoeker did not claim that he had seen anything like a little man in a sperm. Rather, he stated that this was what he thought

Hartsoeker's 1694 drawing of what he thought he would be able to see in a spermatic animal, if he had a sufficiently powerful microscope.

he ought to be able to see, if he had a powerful enough microscope. In other words, his drawing was not a description but a hypothesis – it is a very clear, condensed version of an extreme and literal view of the idea of preformation, produced by one of the less able thinkers of the period. It was never intended to be an actual description of what had been seen, and Hartsoeker's readers were well aware of this difference.[16] Nor was it taken particularly seriously at the time. Given the scepticism that had greeted Kerckring's ludicrous jelly babies, which were allegedly three-day-old embryos, no one would have believed an even smaller version, hunched up in the head of a spermatic animalcule. Strikingly, not even Hartsoeker was really convinced; within fifteen years he had effectively abandoned preformation as an explanation of generation.[17]

More ridiculous, and perhaps deliberately so, were some figures produced by one François de Plantade, a young wag from Montpellier who apparently decided to play a prank on European intellectuals.[18] In spring 1699, Plantade submitted a brief report of his 'investigations' into the nature of semen, in Latin, to the Amsterdam-based journal *Les Nouvelles de la République des Lettres*.[19] The journal published his account – seemingly in good faith – accompanied by a prudish note explaining: 'We leave this Letter in the language in which it was written, because it contains material that cannot be dealt with in French.'[20] Almost simultaneously the same letter also appeared in publications in London and Edinburgh, suggesting that it was a deliberate spoof or hoax.

Under the pseudonym 'Dalenpatius' (a Latinised anagram of Plantade's name), the article explained how the author had seen individual animalcules in the semen which 'scarcely exceed in size a grain of corn'.[21] This gross overestimation of the size of the seminal animalcules should have warned readers that all was not as it seemed. What followed must have clinched it: Dalenpatius claimed that one of the animalcules 'sloughed off the skin in which it had been enclosed, and clearly revealed, free from covering, both its shins, its legs, its breast, and two arms'.[22] In other words, inside the creature was something looking remarkably like a tiny, fully formed, human being. To ram home his satirical point, he included a couple of drawings which showed the 'spermatic animalcules' as little men wearing wiggly hats. If you look closely at the head with its 'hat', you can see that this resembles a sperm, as does the outline of the body and the line between the legs. Dalenpatius clearly seems to have been making fun.

Leeuwenhoek did not see the joke (to be fair, it was not very funny), and immediately wrote a riposte to the Royal Society, which was published in the *Philosophical Transactions*. In his critique, he first asked his readers to compare his findings with the preposterous claims of Dalenpatius ('I believe that no Member of the Royal Society will allow of the discovery of such a Creature, but rather take it to be a Fancy or imagination, then [sic] a real truth'[23]). He also scornfully dismissed all of Dalenpatius's claims; above all, he argued, the idea that the seminal animalcule contained a perfectly formed human being went against every piece of evidence relating to the

1699 drawing by 'Dalenpatius' of a little man inside a spermatic animal, as it appeared in Les Nouvelles de la République des Lettres. *It was probably a joke.*

growth of organisms. As Leeuwenhoek rightly pointed out, embryos do not simply grow larger, they also change their proportions. Even if there was a tiny human in a sperm, it would not look like a little man. Leeuwenhoek's conclusion was forthright and honest, outlining both his conviction that the future adult must in some way be represented in the sperm, and the frontiers of his ignorance: 'I put this down as a certain truth, that the shape of a Human Body is included in an Animal of the Masculine Seed,' he said, before explaining his doubts about the possibility of ever studying the question: 'that a Mans Reason shall dive or penetrate into this Mistery so far, that in the Anatomizing of one of these Animals of the Masculine Seed, we should be able to see or discover, the intire shape of a Human Body, I cannot comprehend'.[24] Leeuwenhoek rightly refused to accept that the something in the sperm that could produce a human was in fact a tiny person.

Despite the fact the Hartsoeker's and Dalenpatius's figures have been reproduced so often since, Leeuwenhoek's robust response shows that neither of these literal and unsophisticated presentations of preformation were at all representative of what most seventeenth-century scientists thought the germinal cells – egg and sperm – actually contained. Thinkers like Leeuwenhoek were generally careful not to be too precise as to what that might be, for the simple reason that they could offer no reliable observations. Nevertheless, they were understandably convinced that because organisms did not literally appear out of thin air, and because development was ordered and regular, something must give rise to that structure and order. It was the only possible explanation.

One of the implicit attractions of preformation was that it avoided the problem of a 'final cause'. With preformation there was no nebulous purpose shaping matter towards an ultimate end – the organism simply grew; it was there from the beginning. Furthermore, as Swammerdam showed when he used 'evaporation' to explain the development of the bee in its pupal case, it was possible to conceive of the kind of growth required by preformation in purely mechanical terms. This was a break with Aristotle, but not with common sense, nor with the spirit of the age. There were two major disadvantages with preformation, however: firstly, the idea was wrong; second, there was virtually no evidence to support it. Nevertheless, neither of these problems, nor the ridiculous claims of Hartsoeker and Dalenpatius, prevented preformation from dominating much of eighteenth-century thinking about generation.

A few thinkers did resist the overwhelming enthusiasm for preformation, but their opposition only revealed more difficulties. In the section dealing with generation in James Drake's *Anthropologia Nova* (1707),[25] the author rightly pointed out that the examples of what we would call blending inheritance argued against preformation in both ovist and spermist versions. He (or she[26]) forcefully dismissed Aristotle's suggestion that the example of the growth of a plant on different soils could explain such phenomena – 'this is so poor, so unphilosophical a shift, that it is not worth an Answer; and they might with as good Authority persuade me, that an Orange-Tree transplanted from *Sevil to England* would bear Apples'.[27] But it was equally clear that the only alternative to preformation was a return to Galen's idea of two semens endowed with some unspecified mystical ability to create life: 'These Difficulties render both these Hypotheses unsatisfactory to me; and however old and exploded the Opinion of a Plastick Power on both sides be, I must however embrace it, even tho' I know not exactly wherein it lies.' Thinkers either had to plump for one of the two partially unsatisfactory theories, or return to the old way of thinking.

Amongst the findings that made the ovist hypothesis of preformation more believable were two contrasting studies of the chicken egg. In 1672, William Croune published a brief description in the *Philosophical Transactions*, in which he claimed to have seen the outline of the future chick in a single unincubated four-day-old egg. His drawing of this phenomenon had about as much value as modern claims to see the face of the Virgin Mary on a slice of toast – it showed a piece of ordinary membrane from which protruded a thin piece of tissue, in which Croune claimed to be able to see 'something very like the head of an embryo of a chick'.[28] Contrasting the

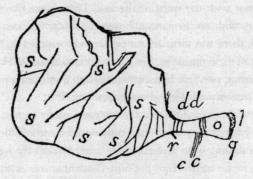

*Croune's 1672 illustration of a 'preformed' chick, as it
appeared in the* Philosophical Transactions. *He claimed
the bit sticking out was the head.*

findings of his single observation with those of the detailed study of
the slow development of the chick described by Harvey, Croune
boldly concluded that Harvey was completely wrong and that 'the
chick is produced complete at one stroke as it were endowed with all
its parts'.[29] Few people accepted Croune's outlandish claim that the
chick appeared suddenly and fully formed, but the fact that it was
published shows the desire to explain embryonic growth, even if it
meant discarding the careful and systematic work of Harvey.

In complete contrast, at about the same time Malpighi sent a
detailed embryological study to the Royal Society. In *De formatione
pulli in ovo* ('On the formation of the chicken in the egg', 1673) he
carefully described all the stages of the development of the chick
embryo, surpassing even Harvey in his accuracy.[30] Apart from the
intrinsic value of his account, there was one aspect of Malpighi's
work which subsequently attracted the attention of eighteenth-century thinkers trying to find support for their highly developed
version of preformation – Malpighi claimed that 'the first outlines of
the principal parts' of the chick could be seen in the egg prior to
incubation, and that 'the first filaments of the chick pre-exist in the
egg and have a deeper origin, exactly as in the eggs of plants'.[31]
Malpighi's ideas were very vague, but he appears to have thought that

the embryo was created fully formed at fertilisation, and then slowly grew and became visible under the microscope. However, he never stated this clearly, nor did he provide evidence to support his idea.[32]

Much of the future success of preformation flowed from a handful of statements by Swammerdam. Although they made up an extremely small part of his work, and often appeared almost as casual asides, they constituted some of his most influential – and to some people's eyes, notorious – findings. Like everyone else, Swammerdam believed that the future organism was somehow contained in the egg. In *Historia Insectorum Generalis* (1669), he noted that a plant seed 'already contains some leaves in its germ, or at least some buds' (try this for yourself – it often works with apple seeds). He then went on to draw an explicit parallel with animals which 'grow from a seed, which contains not only all the parts of the animal, but which is effectively the animal itself, enclosed in a membrane'.[33]

The problem Swammerdam was grappling with was that if you actually looked at an egg shortly after fertilisation it was absolutely featureless. You could look as hard as you wanted for 'all the parts of the animal', but you would not see them – no matter what Croune might have claimed. Harvey's solution had been to talk about 'potential' parts existing in the egg; Swammerdam gave a similar answer, stating that organisms grew 'from invisible, but really existing principles'.[34] Given that much of Swammerdam's work was devoted to showing the changes in the shape and proportions of organisms as they developed, it is clear that he did not believe there was a little fully formed organism in the egg which simply had to be blown up to full size. Indeed, he argued that the individual grew by using material from the egg, and claimed that growth took place 'insensibly and by degrees perfected by the evident external addition of matter'.[35]

The decisive part of Swammerdam's contribution to the later development of preformation came in a throwaway remark in *Miraculum Naturae*, in which he raised the possibility that not simply the immediate offspring was contained in an egg, but, equally, the offspring's offspring, and so on, back to the beginning of the world, indeed to the Garden of Eden, like thousands of Russian dolls. He framed this comment within a reminder of his fundamental belief that all organisms came from a material source ('propagation'), and were not magically generated: 'In nature there is no generation, but only propagation, the growth of parts. Thus original sin is explained, for all men were contained in the organs of Adam and Eve. When the stock of eggs is finished, the human race will cease to exist.' Given that Swammerdam is often thought to have been an ovist, it is striking that he refers here to *both* Adam and Eve. Indeed, a few lines before this example, he used another, more obscure biblical reference (Hebrews 7:1–10) to back up his view of preformation, explicitly suggesting that the future offspring was contained in the *male* parent: 'Furthermore, we can now understand how Levi, being still in his father's loins, paid his tithes before he was born: "For he was yet in the loins of his father, when Melchisedec met him."' All this underlines that Swammerdam's ideas about the question were far from coherent, presumably because he did not consider it to be particularly important.[36]

This extreme version of preformation, known as 'pre-existence', appears to have been suggested to Swammerdam by the Catholic priest and philosopher Malebranche, who presented the latest scientific findings, in particular the ideas of Descartes, in a Christian context. He outlined his ideas in a multivolume work, *Recherche de la vérité* ('Search for the truth'), which became enormously influential as a summary of contemporary ideas about life, the Universe and everything. Malebranche joyously embraced the idea of preformation and took it to the bizarre but logical conclusion of pre-existence.[37]

While pre-existence seems pretty crazy to modern eyes, it was in fact entirely in keeping with the apparently limitless nature of the Universe which was unfolding before the amazed eyes of seventeenth-century thinkers. The mystical French mathematician Blaise Pascal had shown that some numerical series went on infinitely, while Leeuwenhoek's discoveries of millions of minute organisms suggested that there was no limit to the size of organised matter. Inorganic matter appeared to show the same effects – when Leeuwenhoek looked at salt crystals, he discovered that even the tiniest grains showed the same distinctive shape.[38]

Interestingly, one of the reasons why Malebranche advocated pre-existence so strongly was also the point at which he parted company with Descartes and his attempts to find mathematical explanations of the natural world. Descartes' mathematical description of the processes involved in development, which was published posthumously and prompted Thévenot to turn to the study of generation, had been a complete failure, provoking John Ray's sniggering dismissal ('ha ha he'[39]). Malebranche wisely differed with Descartes on this key point. Describing his master's endeavour as 'a little rash',[40] Malebranche explicitly criticised mechanical attempts to describe development: 'It is impossible for the union of the two sexes to produce a work as admirable as the body of an animal. One can well believe that the general laws of the communication of movement are sufficient to form and develop the parts of organised bodies, but it is impossible to believe that they ever form such a complex machine.'[41] In other words, because it was impossible to imagine what laws might under-pin the creation and development of organisms, the only logical solution was pre-existence, which effectively put the appearance of all forms of life back into the lap of God.[42]

Although it is tempting to laugh at pre-existence and preforma-tion, there are striking parallels between the ideas underlying these theories and those that dominate part of modern thinking about genetics. The fundamental idea behind preformation was that the adult organism was contained in the egg, and that development was merely an unfolding of something that had already been decided. Pre-existence simply took this a step further by suggesting that all subsequent generations were also predetermined, encased in either the ovaries or the testicles.

Although everyone would now agree that this was rubbish, part of the idea still has some currency. Barely a week goes by without the media announcing that scientists have discovered the 'gene for' homosexuality, breast cancer, alcoholism, schizophrenia or even shopping. Genes are taken to determine completely our existence, from our size to our health to our habits, acting as directly and ineluctably as the unfolding of an embryo according to the advocates of preformation. 'You cannot escape destiny' is the message. Furthermore, because genes are the units of inheritance, the implication is that these effects will continue down the generations. In a strange way, therefore, a key aspect of preformation and pre-existence lives on in the overstated claims of genetic determinism. This simplistic understanding of the role of genes is shared by those scientists who do not realise the complexity of living systems, or who underestimate the essential role of interactions with the environment, which affect every aspect of the activity and growth of organisms. Twenty-first-century geneticists who suggest that there is a straight line running from a single DNA sequence to a complex human behavioural trait, or even to something as simple as a finger, would be amazed to know that they are the modern equivalents of the seventeenth- and eighteenth-century thinkers who advocated a rigid and literal form of preformation.[43]

In the years between the ground-breaking discoveries of the generation network and the present day, scientists and thinkers continued to study the problem of where organisms come from, and how an egg turns into a fully formed individual. Progress was anything but linear and gradual: there were long periods in which no notable advances were made, others in which people adopted what now appears to be a mistaken approach, and brief, exciting moments when knowledge convulsed, new facts were discovered and, above all, new theories were developed which altered our understanding of how the natural world functions.

Redi and Swammerdam's demonstration that all animals come from an egg laid by a female of the same species was of decisive importance in establishing the 'fixity' of species, itself a key step in the development of our understanding of the natural world.[44] 'Like breeds like' showed that species are normally separate, although it took some time for this idea to be fully accepted; in the 1720s, English people who should have known better swallowed the story of

Mary Toft of Godalming, who allegedly gave birth to rabbits. The certainty that like breeds like gave rise first to the classificatory system of Carl Linnaeus (1707–78) and then the development of various ideas of the evolution of species over time, culminating in the nineteenth century with Darwin's theory of evolution by natural selection.

However, in the decades that followed the work of the Leiden group on the egg, and Leeuwenhoek's momentous discovery of the spermatic animalcules, most thinking about generation focused on the question of where the organism came from.[45] Although spermism enjoyed some popular success, finding ribald expression in Laurence Sterne's baroque novel *The Life and Opinions of Tristram Shandy, Gentleman* (1759–67) – in the opening chapter the hero recounts his early life as a homunculus – by the second half of the eighteenth century the ovist version of preformation prevailed amongst thinkers. One of the reasons for this domination was the discovery of parthenogenesis – generation by females alone, without the involvement of males. Leeuwenhoek and the French polymath René-Antoine Ferchault de Réaumur (1683–1757) had both noticed that aphids gave birth to live offspring, but apparently had no male forms. In the 1740s, the Swiss naturalist Charles Bonnet (1720–93) was able to produce up to ten generations of aphids without the involvement of any males. Parthenogenesis undermined support for the spermist hypothesis and showed that not all generation involved two sexes, but it was clearly limited to a small set of special cases.

In fact, during the eighteenth century it became increasingly obvious that sexual reproduction could be found everywhere, including in plants. This not only provided the basis of much of the new plant classification pioneered by Linnaeus – another Leiden medical graduate – it also resolved the question of exactly what a seed was – i.e. an embryonic plant, or a fertilised 'egg'. However, as the century progressed, thinkers became increasingly impressed by the fact that many plants seemed to be qualitatively different from animals, and did not generate through sexual reproduction, but rather through other means, including budding. The gulf between plants and animals was apparently bridged by a discovery made by one of Bonnet's relatives, Abraham Trembley (1710–84), who observed another organism that had initially been studied by Leeuwenhoek, the freshwater polyp (now called hydra). This odd organism not only seemed to be part-plant, part-animal (hydra are green, yet can move, react to stimuli and catch prey – they are in fact animals, related to corals), it also showed the strangest abilities. It reproduced by budding, and if one of its limbs was chopped off, not only did the animal grow a new limb, but the chopped-off limb would turn into a new individual. In this case, generation clearly did not involve the growth of a preformed hydra in either an egg or a spermatic animal – new hydra could appear from any part of the animal's tissue. This supported the idea of 'pangenesis' – the existence of generative particles in every tissue of the organism, as suggested by the Greek atomist philosophers and advocated by Nathaniel Highmore in 1651.

But while pangenesis might account for the specific findings from hydra, it again raised the problem of how these particles came together to build the organism during generation, with the unfortunate implication that a 'final cause' might be involved. The French naturalist Georges-Louis Leclerc, Comte de Buffon (1707–88), accepted this possibility and argued that organic forms appeared through a process similar to Harvey's 'epigenesis', in which matter was shaped by what he called an 'interior mould'. In general, however, this kind of appeal to unknown forces was even less palatable to mid-eighteenth-century thinkers, influenced by Enlightenment materialist philosophy, than it had been for their seventeenth-century counterparts. Buffon also tried to dismiss the idea of pre-existence by calculating quite how minute the 'Russian dolls' would have to be, but this was not seen as being a decisive criticism at the time – it was not yet clear that there was a lower limit to the size of organised matter.

Preformation was also challenged by a revival of interest in patterns of inheritance, which were intensively studied both by Réaumur and by Pierre Louis Moreau de Maupertuis (1698–1759).

In 1745, Maupertuis, who was deeply opposed to all types of pre-formation, published a book in which he used the example of a 'white negro' (an albino child born to black parents, who had pro-voked a great deal of voyeuristic excitement in Parisian society) as a starting point for trying to understand how characters were inher-ited. Maupertuis suggested that both parents contributed 'particles' to the offspring; if too many of these particles were present, there was more than one offspring, and if there were too many or too few of a certain kind of particle, a monster could be born (monsters were a particular problem for preformation, as it was hard to explain how a healthy parent could give rise to a deformed offspring). In the case of the 'white negro', Maupertuis perceptively suggested that the child's colour could be due to the fusion of 'particles' representing the condition of an ancestor, which were not seen in either parent, or due to some novel change in the nature of the particles. Simply using logic, Maupertuis was groping towards something like the right answer.

In the 1750s both Maupertuis and Réaumur published accounts of the inheritance of polydactyly – having hands with six digits – each using data from a different family, one from Germany, the other from Malta (Maupertuis had evidence from four generations, Réaumur from three). In both cases, it was clear that the character was indeed inherited, and that the pattern of inheritance could not be explained by either version of preformation. Maupertuis even cal-culated the odds that the observed distribution was due to chance – his estimate was that it was something like 8,000,000,000,000:1. However, Maupertuis's work was largely ignored at the time – because he thought that the hereditary particles were distributed throughout the body, he denied that the ovarian follicles had any-thing to do with mammalian generation, and argued that the role of the seminal animals was merely to agitate the generative particles within the semen. Maupertuis was very much in a minority amongst mid-eighteenth-century thinkers: the weight of evidence, while contradictory, increasingly convinced people that ovist preformation was correct, although everyone remained hazy about why both semen and eggs were generally necessary for generation to occur.

One of the few people to attack this problem experimentally was the Italian priest Lazzaro Spallanzani (1729–99),[46] who carried out a series of experiments on the role of semen in the generation of frogs, in which fertilisation takes place externally. To prove his hypothesis that contact between egg and semen was necessary for generation to occur, Spallanzani carried out a surreal experiment in which he dressed his male frogs in little taffeta shorts. Semen released by frogs wearing the shorts could not come into contact with the eggs, and no tadpoles were produced. However, if the eggs were touched with a paintbrush that had been dipped in frog semen, they developed into tadpoles. Having also carried out artificial insemination in dogs, toads and silkmoths (succeeding where Malpighi had failed), Spallanzani went on to study how the ability of semen to fertilise was affected by dilution, desiccation, filtration and heating.

Although Spallanzani's findings clearly showed that there was something very small in the semen which had the ability to induce growth by contact with the egg, and which could be disrupted by heating, and the same was true of the seminal animals, Spallanzani remained a convinced ovist. The strength of the preformation argument and the difficulties associated with epigenesis, together with the lack of any alternative theory, led him to argue that semen did nothing more than provoke the activity of the foetal heart, while the hundreds of millions of tiny spermatic animals were merely parasites that lived in the male testicles and were passed from father to son at the moment of generation. Spallanzani's obstinate focus on ovism was part of a strange paradox: although everyone could see the spermatic animals, no one had actually seen the egg in a viviparous organism. As the twentieth-century philosopher of science Jacques Roger put it: 'In the middle of the eighteenth century, when almost everyone believed in the existence of eggs of vivipara, which no one had seen, no one knew what to do with the spermatic animalcules which everyone could observe.'[47]

Despite all the work that had been done since the 1660s, there was still no agreement about how generation worked. The general uncertainty was summed up at the time by Professor Wristberg of Göttingen University, who listed all the contributions to the study of generation, from Harvey and De Graaf to Spallanzani and Trembley, and sighed: 'all these I say, have convinced me, that our whole knowledge of generation consists merely of phenomena, upon which it has as yet been impossible to raise a true and infallible system'.[48]

During the nineteenth century important steps were taken towards such a 'true and infallible system', as new ideas and new technologies helped scientists begin to understand the roles of eggs and sperm, their true nature, and the way they produced a new individual. But as a result of these findings, the concept of 'generation' began first to fragment and then to disappear. The very word started to be systematically replaced by 'reproduction', the more precise term adopted by Buffon in 1749, which implied that like bred like, and also tended to exclude the 'development' aspect of 'generation', focusing on the mechanisms of fertilisation and, eventually, inheritance.

The first of these increasingly separate aspects of generation to be understood was also the most fundamental: the roles of egg and sperm in conception. Remarkably, very little progress had been made since the time of De Graaf and Leeuwenhoek. In the 1680s Malpighi had described the empty ovarian follicle in the cow as a glandular *corpus luteum* ('yellow body'), and in 1691 Nuck found evidence from dogs which he thought proved that embryos came from fertilised eggs, rather than from the male's semen, but for much of the following century there had been few investigations into the nature of mammalian conception.[49] In the final decades of the eighteenth century, attention began to turn back to the ovaries. In Scotland the Hunter brothers looked at the effects of removing one ovary, and in 1797 two replications of De Graaf's work on rabbit reproduction were published in the *Philosophical Transactions*.[50] But none of these studies provided a decisive breakthrough in thinking about generation. The first sign that things were about to change came between 1824 and 1827, in a series of articles by two young

French scientists, Jean Prévost (1790–1850) and Jean Dumas (1800–84). First, they developed Spallanzani's work and found strong circumstantial evidence from frogs that each egg was fertilised by a single spermatic animal. They then showed that if the spermatic animals were killed, the semen lost its fertilising power. Finally, they turned their attention to rabbits, repeating De Graaf's observations, clearly stating that the ovarian vesicles must release a tiny egg – they even described a small structure within the vesicle just prior to rupture which was presumably the egg or oocyte, although they did not identify it as such.

In the same year as Prévost and Dumas published their final article, the young Estonian physician Karl Ernst von Baer (1792–1876) made the first description of the mammalian egg.[51] Having been primed by the work of Prévost and Dumas, Von Baer knew what he was looking for, and went searching for the egg in the ovarian vesicle of a dog. Nonetheless the result was still a huge surprise, as he later recalled: 'I opened one of these little sacs, lifting it carefully with a knife on to a watchglass filled with water, and put it under the microscope. I shrank back as if struck by lightning, for I clearly saw a minuscule and well-developed yellow sphere of yolk. Before I found courage to look at it a second time, I had to recover, since I was afraid of having been deluded by a phantom. It seems odd that a sight expected and so much longed for could frighten one when it actually occurs.'[52] Von Baer went on to compare the mammalian oocyte with eggs from a variety of other species, including frogs, chickens and crayfish, providing a series of drawings showing the egg within the follicle, ruptured follicles, and so on. His conclusion was simple: 'Every animal which springs from the coition of a male and female is developed from an ovum.'[53] While this was not quite as pithy as the 'Ex ovo omnia' of Harvey's frontispiece, nor as revolutionary as Swammerdam's 'all insects proceed from an egg, that is laid by an insect of the same species', it sounded the death knell of spermism.

Like everyone else, Von Baer recognised that both egg and semen were necessary for conception to occur. But, despite having finally identified the female component, he remained a resolute ovist. Like Harvey, he argued that something in the semen – but certainly not the spermatic 'animals' – merely awoke the egg. Von Baer's rejection of any role for the seminal animals flowed from his conviction that they were parasites, a kind of worm from the genus *Cercariae*, which included the flukes. He even gave the spermatic animals a name – 'spermatozoa' (literally: animals of the semen), by which they are still known today, even though we now realise that they are not 'animals' at all.

Once again, scientific history seems to have stood still: on the brink of drawing what appears with hindsight to have been an obvious conclusion, thinkers did not realise what was staring them in the face. This was precisely the problem. Despite 150 years of research, no one had any real idea what egg and spermatozoa were, what they represented, or what they contained. From this point of view, generation remained as great a mystery at the end of the 1830s as it had been in the 1670s. To reshuffle existing knowledge into something resembling the truth, what was needed was neither more microscopic investigations, nor more frogs in shorts, but simply a different way of looking at what was already known.

The key conceptual breakthrough occurred when the development of the 'cell theory' led to a complete re-evaluation of many existing ideas about the natural world.[54] In 1838 and 1839, two German biologists, Matthias Schleiden and Theodor Schwann, finally brought together the casual analogy made by Robert Hooke in *Micrographia*, and the observations of the eighteenth-century embryologist Caspar Friedrich Wolff, who had suggested that all animal and plant life was made up of 'globules'. Schwann and Schleiden put forward the hypothesis that all life is made up of cells, and specifically argued that the animal egg, like the plant embryo, was a cell. The history of the 'cell theory' was complex and is still hotly debated today, but the outcome was that 'biology', the new term for the study of life, was able to focus on a fundamental unit that was equivalent to the elements that had been discovered in chemistry. This approach encouraged scientists to look at life in terms of its simplest components, rather than as a complex whole.

The cell theory also provided a very powerful explanation for what exactly the spermatozoa and egg were. If they too were cells, that not only suggested how the two components were equivalents, despite their huge differences in size and motility, it also hinted at what they might do during conception – fuse to make a new cell. By 1853 it had been shown that during fertilisation a single spermatozoon actually entered the egg (something Leeuwenhoek had suggested in 1683) and at the end of the 1870s it became clear that the egg and sperm cells contributed equally to the creation of the embryonic cell.[55] With these discoveries, which were also extended to plants, the essential mechanics of the conception aspect of generation had been described. But generation as a whole was still not understood – the twin questions of what was in the two cells, and how they created a new organism, remained unanswered.

As well as discovering the mammalian egg, Von Baer also laid the bases for modern embryology – the study of the development of the organism, from fertilised egg to birth. This science, which was born in the seventeenth century with the work of Fabricius, Harvey, Swammerdam and Malpighi, came to full fruition in the second half of the nineteenth as the focus shifted from theoretical explanations to increasingly accurate descriptions of what could be seen under the microscope. The huge developments in embryology during the nineteenth century were driven by major technical developments in optics, by the invention of new chemical stains which highlighted bits of organs and cells that were previously invisible, and by the study of new experimental organisms, such as the sea urchin, which enabled scientists to observe conception in unprecedented detail. This new experimental embryology put an end to the crude preformationist view of growth as simply involving the unpacking of tissues. Instead of containing a tiny preformed organism, the embryo was revealed to be a complex set of folded and pleated structures, with development involving the distorting effects of waves of contraction and expansion in different parts of the body. As scientists watched the strange ballet of stretching and bulging tissues unfolding under their shiny brass microscopes, they were finally able to answer the question that had so perplexed their seventeenth-century equivalents, and describe how new tissues grow.

A key part of the cell-centred vision of life that came to dominate biology from the middle of the nineteenth century was the notion that all cells come from another cell or, as the German pathologist Rudolf Virchow (1821–1902) put it in 1855, '*omnis cellula e cellula*' ('every cell comes from a cell'). Cell division and cell death are the essential forces that sculpt the growing organism, allowing some bits to get bigger, and others to shrink. Through the findings of nineteenth-century embryology, Swammerdam's assertion that development simply involved 'propagation and the growth of parts' was both confirmed and made to look a bit leaden. While these are indeed the key events that take place, development turns out to be an incredibly rich process which we still only barely understand.

Von Baer had noticed that the embryo became increasingly complex as it grew, going from the general – a common embryonic structure – towards the specific – the final form of the given organism. After the publication of Darwin's *The Origin of Species* in 1859, the German zoologist Ernst Haeckel provided an explanation of this observation, summed up in his strange aphorism 'ontogeny recapitulates phylogeny'. This simply meant that Haeckel thought that during development ('ontogeny') the embryo goes through all the adult forms of the evolutionary ancestors of the species.[56] Hugely influential in its time, Haeckel's idea was completely wrong. Although the initial stages of all vertebrate embryos are indeed similar, they do not resemble ancient adult organisms, and look alike only because all vertebrates share basic structures, which are made using a set of genes that goes deep into evolutionary history. By the early decades of the twentieth century, Haeckel's theory had been abandoned. But despite being mistaken, it had given an intriguing answer to the question of where embryonic order came from. Epigenesis explained development by harking back to Aristotle's 'final cause', while extreme preformationists had thought embryonic structures were

simply packed away, perhaps going back to the Creation. The implication of Haeckel's view was that the order that could be seen during
development came from the evolutionary past. Understandably
enough, Haeckel had no idea how that order could be expressed
across millions of years, but the explanation would soon be found.

The final piece in the generation jigsaw was the appearance of the
science of genetics in the early years of the twentieth century.[57]
Geneticists rightly trace their science back to the work of Gregor
Mendel, whose studies of inheritance in pea plants were published to
resounding indifference in 1866, only to be rediscovered in 1900.
Mendel solved the problem of varying patterns of inheritance which
had so perplexed Harvey and others by suggesting that heredity
involved the action of 'particles', the effects of which could be
observed by studying large data sets and using simple statistics. In
itself, this was an amazing insight, but it was not modern genetics.
Mendel's views were not simply rediscovered in 1900; they were also
reinterpreted in the light of new ideas developed in the second half
of the nineteenth century, which had been synthesised by the
German zoologist August Weismann (1834–1914). Weismann argued
that inheritance was based on recently revealed structures in the
cell – the chromosomes – which were transmitted from parent to
offspring through the 'germ cells' (egg and sperm), and which were
not affected by the 'somatic cells' (the rest of the body). This framework, which was applied to both animals and plants, suggested that
inherited characters could not be acquired, and that the units of
inheritance which determined how organisms developed were
located on the chromosomes.

By the 1920s the new science of genetics, as it was termed, was
triumphant. Following the work of Thomas Hunt Morgan and his
young collaborators on the tiny fruit fly Drosophila, the shape of
human knowledge about generation was altered for ever. It was now
clear that in most animals and many plants, each individual has two
copies of each gene, one copy on each of a pair of chromosomes,
with the number of chromosomes varying enormously between
species – fruit flies have four different pairs, while humans have

twenty-three different pairs (the situation can be more complicated in plants). During the creation of sperm (or pollen) and egg, the chromosomes are shuffled and rearranged, with only one chromosome of each type ending up in each germ cell. Egg and sperm therefore contain half the genes that are present in the parent organism; at conception the two germ cells fuse, creating a new cell with the full complement of genes. This model allowed scientists to understand the various kinds of inheritance – although most characters were determined by many genes, and therefore show blending inheritance, others are determined by a single 'dominant' character, while in many animals sex is determined by the presence of a particular chromosome, and therefore shows no blending.

The final step in resolving the mystery of reproduction involved the unravelling of the genetic code in the 1950s and 1960s, following James Watson and Francis Crick's momentous discovery in 1953 of the double-helix structure of the DNA molecule which makes up the chromosomes. Watson and Crick not only discovered what the DNA molecule looks like, they also revealed how it can copy itself and therefore how it can transmit information from one cell to another and from one generation to another. New concepts, taken from computing, shed light on the processes involved in turning genes into proteins, which in turn are used in cellular processes and in building organisms and allowing them to function.

The rise of genetics led to the death of generation as a biological concept. The study of the appearance of new life became separated into a series of subjects – reproductive biology, developmental biology and genetics. The separation of genetics and developmental biology meant that for much of the twentieth century scientists had only the haziest ideas of the precise mechanisms underlying differentiation – how the single-celled zygote turns into a screaming, wriggling baby. Over the last twenty years, however, following the work of 1995 Nobel Prize winners Ed Lewis, Christiane Nuesslein-Volhard and Eric Wieschaus on the genetics of development in *Drosophila*, there has been a resurgence of interest in how development occurs, its genetic bases and its role in evolution. Understanding the nature and evolution of what Swammerdam called 'the rules and theorems of generation' has become one of the most exciting challenges of twenty-first-century science.

At the same time, huge advances in reproductive biology have led to a radical reworking of all the old ideas about generation. Probably the greatest impact of studies of generation on everyday life has come through the discoveries associated with the hormonal mechanisms of reproduction, and the development of the contraceptive pill by Carl Djerassi in the early 1950s. Curiously enough, one of the ultimate fruits of the study of generation has been the widespread availability of the means to stop it happening. We can now also assist those who want to have children but cannot. In 1978 the world's first 'test-tube baby', Louise Brown, was born in the UK. Since then, *in vitro* fertilisation (IVF) has become routine, and around two million people are currently walking around who were conceived in a test tube, rather than by the more usual, and pleasurable, route.

In 1996, cloning technology, which was already well established in amphibians, finally succeeded in mammals with the birth of Dolly the sheep, who was soon followed by a growing menagerie of other animals. The nucleus of an egg was replaced with the contents of an ordinary cell, producing life without the intervention of sperm. The looming possibility remains that someone will eventually clone a human being, with all the ethical problems and potential health implications for the clone that such a step would inevitably involve. Ironically, many of the preconceptions about cloning hark back to preformation – the unlikely suggestion that any clone would be identical in every respect to its original, down to its personality – while debates about limits on the use of human stem cells can resemble medieval discussions about 'ensoulment' – the moment at which the 'soul' was supposed to enter the embryo. Generation may be gone from the laboratory, but some of the old ideas associated with it continue to reverberate in both scientific debate and the popular imagination.

About the only context in which 'generation' is still used in biology is in historical discussions of spontaneous generation. Leeuwenhoek's discovery of micro-organisms raised the possibility, which even he initially accepted, that these 'animalcules' might be an exception to what appeared to be the general law that like breeds like. It was possible, some people felt, that micro-organisms could generate spontaneously. In the eighteenth century, Spallanzani clashed with John Turberville Needham (1713–81) over this question. Having found that water infused with vegetable matter that had been heated and sealed nevertheless produced micro-organisms, Needham claimed that all organic matter had a generative power. Spallanzani disagreed, and carried out a series of experiments which suggested that Needham's findings could be explained by insufficient heating, or contamination by existing micro-organisms.

This debate continued into the second half of the nineteenth century, as argument about the spontaneous generation of micro-organisms and of internal parasites[58] culminated between 1858 and 1863 in the confrontation of two giants of French science, Félix Pouchet (1800–72) and Louis Pasteur (1822–95). Pasteur ferociously opposed Pouchet's claims to have found evidence in favour of spontaneous generation, and in a series of debates and experiments involving his 'swan-necked' flasks, succeeded in convincing the French Académie des Sciences that he was correct (as indeed he was). That did not mark the end of the issue, however, and in the 1870s the English physician Henry Bastian (1837–1915) claimed to have new evidence in favour of the hypothesis. But despite the support of those in the medical community who defended the idea of the spontaneous generation of disease-causing microbes, most scientists were not convinced by Bastian's arguments. From the beginning of the twentieth century, everyone was satisfied that no living organisms are spontaneously generated.

Spontaneous generation is not all nonsense, however. After all, the first organism had to come from somewhere – at some point, long ago, matter must have turned into life. Unless we accept either the lazy argument of 'panspermia' – the idea that life came from outer space, which simply pushes the problem back to another planet – or the even lazier idea of divine intervention, the only solution is spontaneous generation deep in the geological past.[59] The exact processes involved are unknown but they were presumably incredibly slow, involving the chance association of organic molecules in small, well-protected, stable and watery environments, such as tiny bubbles in warm mud. These processes may even be continuing today, although if this were the case natural selection would operate with a vengeance, and any half-formed organic structure would almost certainly be eaten by pre-existing life forms before it got the chance to evolve any further.

Some of the many schemas which account for the generation of life on Earth have been experimentally tested, and the results have supported the idea that matter can spontaneously self-organise. No one has yet created life in a test tube; the original process which led to life as we know it probably took millions of years, in a 'laboratory' the size of the planet. These ideas are currently the subject of a great deal of speculation, not least because they raise the possibility of focusing the search for extraterrestrial life on planets, moons, asteroids or comets that are likely to support the appropriate conditions. 'Generation', dead for around a century as a scientific concept, may yet return in order to explain the origin not only of life on Earth, but also of little green men.

Although the seventeenth-century thinkers who investigated generation studied a wide range of organisms, from flies to frogs and from deer to dogs, the subject that ultimately preoccupied them all was the generation of human beings, and in particular the reproductive power of women. Despite the fact that their discoveries had little immediate effect on the lives of ordinary people – ignorance about conception, inheritance and growth was widespread until the middle of the twentieth century, and still persists in many places – that does not mean they were unimportant. Eventually, their work led to the changes in our understanding of reproduction which gave women the ability to control their fertility through effective contraception, thereby changing their role in a way which has altered everyday life around the world. By overcoming common-sense folk beliefs in spontaneous generation, these thinkers laid the basis for our modern view of disease, providing humanity with the means to reduce suffering and even eradicate maladies which in the past had killed millions. Even our realisation that life has evolved through natural selection can trace its origins back to the seventeenth-century demonstration that like breeds like, and that generation, in all its various aspects, is lawful.

Today most of the men who made these discoveries have been forgotten by everyone except historians of science, while the full details of their work are known to only a handful of people. But these men and their story deserve to be remembered. Without their work, without their insights and without their productive errors, the history of science – the history of ourselves – would be poorer, and very different. They shaped the way human knowledge developed, proving the power of experimentation and providing the basic answers to the question that had preoccupied humanity since the dawn of time. They showed us where we come from.

CHRONOLOGY

1651   Harvey: *Exercitationes de Generatione Animalium*. Harvey's book popularises the idea that all organisms come from an egg, but fails to provide any clear evidence

1663   Steno, Swammerdam and De Graaf meet in Leiden

1664   Steno and Swammerdam visit Paris and come under the influence of Melchisedec Thévenot

1665   Swammerdam dissects a silkworm and shows that the structures of the adult moth can be seen within it
De Graaf arrives in Paris; Swammerdam and Steno leave
Kircher: *Mundus Subterraneus*. Kircher gives various examples of how to generate insects from decay

1666   Swammerdam shows that gall insects come from an egg laid by a female insect
Steno arrives in Italy and meets Francesco Redi

1667   Steno: *Canis Carchariae Dissectum Caput*. In a throwaway phrase, Steno suggests that women have eggs

1668   De Graaf: *De Vivorum Organis Generationi Inservientibus*. De Graaf's book describes the male genitalia in humans and promises an equivalent study of women
Van Horne publishes a letter claiming to have been the first person to say that human ovaries contain eggs; this leads to a row with De Graaf over who had first described the structure of the male testicle

1668 Redi: *Esperienze intorno alla generazione degl'insetti*. Using careful experimentation, Redi proves that insects are not spontaneously generated, but he remains uncertain about gall insects

1668–9 Feud between De Graaf and the Royal Society over priority in describing the structure of the male testicle

1669 Malpighi: *Dissertatio Epistolica de Bombyce*. In the first study of an insect's internal anatomy, Malpighi shows that caterpillar and butterfly are the same organism, and attempts artificial insemination of moth eggs

  Swammerdam: *Historia Insectorum Generalis*. Swammerdam states that all organisms come from an egg laid by a female of the same species, and shows that all growth is slow and ordered

1670 Van Horne dies without publishing any further material on generation

1671 Kerckring: *Anthropogeniae Ichnographia, sive Conformatio Foetus ab Ovo*. Kerckring shows unbelievable pictures of early embryos as little jelly babies. De Graaf and Swammerdam realise that they risk being upstaged in their respective claims over the human egg

1671 Swammerdam's illustration of the uterus and ovaries is sent to the Royal Society to stake his claim for priority over De Graaf

1672 De Graaf: *De Mulierum Organis Generationi Inservientibus Tractatus Novus*. Using careful experimentation on rabbits, De Graaf provides compelling evidence that women's ovarian follicles release eggs, but does not actually observe the egg

  Swammerdam: *Miraculum Naturae, sive Uteri Muliebris Fabrica*. Swammerdam outlines his and Van Horne's claim to have been the first to observe the human 'egg', and is highly critical of De Graaf's work

  Swammerdam sends a preserved uterus and ovaries to the Royal Society, as further proof of his findings.

1673 De Graaf: *Partium Genitalium Defensio*. De Graaf's blistering polemical reply to Swammerdam's claims provides little extra evidence, but makes many allegations

  De Graaf dies. Royal Society reports on priority dispute between Swammerdam and De Graaf, giving the credit to Steno

1677	Leeuwenhoek's letter to the Royal Society reports existence of spermatozoa
1680	Swammerdam dies
	Thinkers divided between 'ovists' and 'spermists', each emphasising one component of generation. Ovists think that sperm are parasites, spermists think that eggs are merely food
1740s	The term 'reproduction' begins to replace 'generation'
1827	Von Baer is the first person to see the human egg, but remains convinced that sperm are parasitic animals
1840s	Rise of 'the cell theory' and realisation that egg and sperm are cells
1860s	Pasteur demonstrates that not even micro-organisms show spontaneous generation
1870s	Fusion of sperm and egg observed
1900	Rediscovery of Mendel's work on pea plant inheritance from 1860s. Beginning of the rise of genetics
1950s	Double-helix structure of DNA and genetic 'code' are revealed
1960s	Contraceptive pill starts to be widely available
1978	Louise Brown, first 'test-tube baby', is born
1995	Nobel Prize for work on the genetics of development
1996	Dolly the sheep is cloned in Edinburgh by inserting the contents of an ordinary cell into an 'empty' egg. Life is generated without the intervention of sperm

NOTES

CHAPTER 1: IN THE BEGINNING (pp. 9–30)

1 Schulz, Van Andel, Sabelis and Mooyaart (1999).
2 See Daston and Park (1998). Even today, these ideas have still not completely disappeared – in June 2004, the BBC News Online website announced: 'Iranian woman "gives birth to frog"'! After a barrage of complaints, the BBC took the page down.
3 Van Helmont (1662), p. 113. In 1864, Van Helmont's recipe was ridiculed by Louis Pasteur as part of the debate over spontaneous generation in the French Académie des Sciences (Pasteur, 1922, p. 329), and has since been repeatedly reproduced, using Pasteur as the source. I expected to find that Van Helmont's claim was a bit more nuanced when it was read in context. It is not.
4 Report by Robert Moray at the meeting of 9 March 1664, in Birch (1756–7), vol. 1, p. 393.
5 Needham (1959), p. 45.
6 For the sources of these and many other tales, see Thompson (1957). These traditions are surely no more bizarre than the miraculous birth celebrated by Christians every 25 December.
7 Gonzalès (1996).
8 Randolph (1638), p. 102.
9 Walton (1653), p. 167.
10 For a discussion of this literature, interpreted in the light of its relation to the changing political context in seventeenth-century England and in particular to views of women, see Fissell (2004). For discussions of French women's understanding of generation, see Duchêne (2004) and Tucker (2004).
11 Raynalde (1654). In previous editions his name was often spelt Raynald. See Fissell (2004).

12 For an account of the situation of many poor women in seventeenth-century England, especially in relation to childbirth, see Gowing (2003). For a more general account of the condition of women at the time, see Hufton (1995).

13 Drake (1707); for a discussion of Drake's work see Cohen (1997).

14 See Holly Tucker's description of the subtle traces of the new science to be found in late seventeenth-century French fairy tales (Tucker, 2004).

15 For a general discussion of the place of the female in Aristotle's biology, see Mayhew (2004).

16 For a discussion of Aristotle's views about spontaneous generation, see Lennox (1982).

17 Sometimes this view had a nasty consequence: a woman who fell pregnant after being raped could be accused of having enjoyed it, and thus having been a consenting partner (Gonzalès, 1996).

18 Galen (1968).

19 For a discussion of Chinese ideas relating to reproduction see Furth (1987, 1999). Things were no better when it came to the generation of animals: for example, because of the massive importance of silkworm farming in China, there were more than enough books written on the care of the silkworm, but there is no sign that any of them tried to explain the processes of generation involved in its life cycle (Bray, 1984, p. 76). For a discussion of a similar lack of interest in generation amongst European silkworm farmers in the seventeenth century, see Chapter 5 below.

20 Cadden (1993) provides a detailed explanation of medieval attitudes to the differences between the sexes, while Orme (2001) describes medieval attitudes to children. Potts and Short (1999) discuss the evolution of ideas about sex and sexuality.

21 Van der Lugt (2004), pp. 462–73.

22 Pagel (1982).

23 This and subsequent quotes are taken from Paracelsus's book *Concerning the Nature of Things* (Jaco, 1995).

24 For a taste of the debates around this period in history, see Cohen (1994) and Shapin (1996). Shapin begins his book with a tongue-in-cheek provocation: 'There was no such thing as the Scientific Revolution, and this is a book about it.' I think there *was* a scientific revolution, and this book shows some of the reasons why.

25 The relation between political and scientific revolution in seventeenth-century England was particularly convoluted: many

of the scientific radicals were enthusiastic Royalists, while the leaders of the most radical wing of the English revolution were inspired not by scientific materialism, but by the mystical aspects of alchemy. For a discussion of the scientific and ideological views of the left wing of the English revolution, see Mulder (1990). There has been a long academic debate about whether there was a connection between science and extreme versions of Protestantism (in particular Puritanism). I agree with Richard Greaves' pithy summary: 'There is a relationship between Puritanism and science, but *not* a direct one. The mediating link is revolution' (Greaves, 1969, p. 368). For an overall account of the complex relations of science, culture, religion and politics in the English revolution, see Mendelsohn (1992) and Rogers (1996).

26 Freedberg (2002).

27 I want to avoid what Katharine Park and Robert A. Nye called 'an anemic and self-congratulatory narrative of scientific discoveries' (Park and Nye, 1991, p. 54), but there is no escaping the fact that the history of science involves discoveries that are often made by individuals, and that much of this story focuses on how we discovered the way generation works.

28 This and subsequent verbatim quotes are from George Ent, 'Epistle Dedicatory', in Harvey (1981), p. 3. Ent does not give the year of this visit, simply referring to 'last Christmas'. It seems unlikely to have been Christmas 1649, as either Ent or Harvey would undoubtedly have mentioned the execution of Charles I, which took place in January 1649. It could have been Christmas 1647, but that would mean a gap of nearly three years before the publication of Harvey's book, which appeared in March 1651. Following Keynes (1966), I have chosen 1648. There is a general problem of dating in this period. In 1582, Pope Gregory XIII decreed that a new calendar (the one we use today) should be used in Christian countries; the old Julian calendar, which went back to Roman times, was gradually getting out of step with the Earth's progress round the Sun. Most European countries adopted the new calendar, but England maintained the 'old style' until 1752. Throughout most of the period covered here, dates in England were ten days behind much of continental Europe. To prevent the text from becoming cluttered with notes about whether a given date is old or new style, I have simply ignored the question, generally giving dates to the nearest month. Where precise dates are

given, they are those that applied in the country concerned.

29 This and subsequent verbatim quotes from George Ent, 'Epistle Dedicatory', in Harvey (1981), pp. 3–5.

30 Harvey's political allegiances and the momentous period he lived through have led a number of historians to analyse his writings with a view to detecting the traces of the English revolution. Most of these studies have focused on his work on the circulation of the blood, *De Motu Cordis* (Hill, 1964, 1965; Whitteridge, 1965; Graham, 1978; Rogers, 1996); but there have also been historico-cultural analyses of one of his dissections (Shepard, 1996), and of *De Generatione* (Keller, 1999).

31 Aubrey (2000), p. 144.

32 Harvey (1981), p. 354. The first edition was published in Latin (Harvey, 1651) and was soon followed by an English translation (Harvey, 1653). The contemporary vocabulary of the 1653 translation has a great atmosphere, but it can be difficult for modern readers to understand. In 1847 Henry Willis published a new translation of *De Generatione*, which is accurate but lacklustre. I have chosen to use the most recent translation by Gwyneth Whitteridge, which is precise and clear for the modern reader.

33 Quotes in this paragraph from George Ent, 'Epistle Dedicatory', in Harvey (1981), p. 5.

34 Shepard (1996).

35 Even a sympathetic biographer has admitted that Harvey's book is 'repetitive, confused and sometimes contradictory'. Keynes (1966), p. 355.

36 Quotes in this paragraph from Harvey (1981), pp. 298, 8, 22.

37 For an analysis of how the aphorism came to be directly associated with Harvey, see Meyer (1936), pp. 79–81.

38 Voltaire, *L'Homme aux quarante écus*, quoted in Cole (1930), p. 150.

39 Harvey (1981), p. 334.

40 Harvey's view of generation has been analysed by Keller (1999) in terms of the prevailing gender relations of the time. Struck by the contrast between Harvey's ambitions and his lack of a clear explanation, Keller argues that his failure to observe semen in the uterus following copulation threatened both contemporary visions of paternity and the patriarchal relations that characterised seventeenth-century England. As a result, she suggests, his interpretation of his findings tended to devalue the role of women, in particular by emphasising the relative independence of the egg.

This argument is slightly undermined by the fact that Harvey's comments about independence refer only to chicken eggs, not humans' (see especially Exercise 26). For a discussion of Harvey's study of generation from a psychoanalytic point of view, see Fischer-Homberger (2001).

41 Harvey (1981), p. 22. There have been a number of attempts to suggest that Harvey did not in fact believe that insects were spontaneously generated. See for example Meyer (1936), pp. 45–55, and Foote (1969).

42 Harvey (1981), p. 189.

43 Ibid. (1981), p. 353.

44 Ibid. (1981), p. 165. For a discussion of female ejaculation, see Blackledge (2003).

45 Ibid. (1981), p. 344.

46 Aristotle, *The History of Animals*, chapter 50.

47 Ibid.

48 Short (1977, 2003). Professor Short's film lasted twenty-five minutes and was entitled 'Harvey's Conception' (Short, 1978). He gave copies of his film to the London libraries of the Wellcome Trust History of Medicine Unit and the Royal College of Physicians (Short, personal communication). He has now generously made it available to the public, and a downloadable copy is available at www.egg-and-sperm.com

49 Harvey (1981), p. 443.

50 Ibid. (1981), p. 448.

51 Ibid. (1981), p. 445.

CHAPTER 2: FRENCH CONNECTIONS (pp. 31–62)

1 De Graaf's first name is sometimes given as Regnier or its Latinised version Regnerus. Steno is known as Nicolaus.

2 The details in the opening paragraphs are taken from Zumthor (1959), Schama (1991) and Israel (1995, 1998).

3 Although Karl Marx was convinced that 'Holland was the model capitalist nation of the seventeenth century' (*Capital*, vol. 1, chapter 31), in the twentieth century Dutch Marxist historians were divided on the question (see Van der Linden, 1997).

4 Engel (1986). For details of Jan Swammerdam the elder's 'cabinet', see the catalogue assembled by Swammerdam after his father's

death (Anonymous, 1679a) and Nordström (1954–5). 'Cabinets of curiosities' are discussed in Pomian (1990); Findlen (1994); Daston and Park (1998); Mauries (2002); Smith and Findlen (2002). A sumptuous facsimile of the catalogue of one of the most famous late seventeenth-century cabinets – that of Albertus Seba – can be found in Seba (2002), together with an introductory essay by Irmgard Müsch.

5 Data collected from Engel (1986).

6 De Monconys (1666), p. 151.

7 Israel (1998), p. 621.

8 Ibid. (1998), p. 572.

9 Lunsingh Scheurleer and Posthumus Meyhes (1975); Kidd and Modlin (1999). For a discussion of the importance of the Leiden amphitheatre, see Rupp (1990).

10 The plans of the amphitheatre have survived, and the Dutch Museum of Science and Medicine, the Museum Boerhaave in Leiden, has reconstructed it, together with the skeletons that surrounded it. A virtual tour can be undertaken from the museum's website www.museumboerhaave.nl

11 Hett (1932).

12 Steno (1661).

13 Kardel (1994a).

14 Nordström (1954–5).

15 Hsia and Van Nierop (2002).

16 Nordström (1954–5).

17 Details in these paragraphs from: Schama (1991); Cook (1994); Dekker (1999, 2001); Roberts (2004); Roberts and Groenendijk (2004); and from the Leiden University Library (see Manuscript Archives, p. 299).

18 Ray (1738), p. 47.

19 Ibid. (1738), p. 28.

20 French (2003), p. 157.

21 There are few studies of Van Horne. The details here are taken from Luyendijk-Elshout (1965). For Sylvius see Wells (1949) and Underwood (1972). The most detailed work on Sylvius is in Dutch (Baumann, 1949).

22 Smith (2004).

23 Porter (1963).

24 Quoted in Luyendijk-Elshout (1965), p. 36. Steno was not the only one to have his doubts. In November 1664 Robert Hooke

copied one of Swammerdam's experiments and found the whole
procedure so upsetting that he refused to repeat it: 'I shall hardly
be induced to make any further trials of this kind, because of the
torture of the creature: but certainly the enquiry would be very
noble, if we could any way find a way so to stupefy the creature,
as that it might not be sensible.' Quoted in 'Espinasse (1956), p.
52. Three years later, the Royal Society insisted that Hooke repeat
the experiment. Clearly unhappy, he twice postponed the ordeal,
then, when the date was set, excused himself. However, the
experiment was botched, and he eventually found himself obliged
to do the job again 'and it succeeded well'. Hooke's final report
can be found in Hook (sic) (1667).

25 De Graaf (1664).
26 Anonymous (1667a).
27 Cole (1944), p. 276.
28 A recent decree from the young King Louis XIV had limited the
    size of each shop sign to one square foot, high up on the narrow
    front of the building.
29 Details in opening paragraphs from De Crousaz-Crétet (1922);
    Howell (1650), p. 25; Lister (1699).
30 For accounts of the early years of the Royal Society and its wide-
    ranging scientific interests, see Jardine (1999) and Gribbin (2005).
    The *Philosophical Transactions* was a simple magazine-format publi-
    cation, each issue of around twenty-four pages, which appeared
    more or less monthly from 1665 onwards, costing sixpence an issue
    (Robinson and Adams, 1935, p. 45). The journal was in fact the
    property of the Secretary of the Royal Society, Henry Oldenburg,
    and after his death in 1677 it appeared much less frequently, not
    appearing at all between 1679 and 1683 (during this time
    Oldenburg's rival, Hooke, irregularly produced his own journal,
    the *Philosophical Collection*). From 1683 the *Philosophical Transactions*
    reappeared; since 1752 it has been published by a committee of the
    Royal Society.
31 Hall (1991), p. 9.
32 The best known of these satirical plays was Shadwell's 1676 pro-
    duction *The Virtuoso* (Shadwell, 1966), which ridiculed the Royal
    Society in general and Hooke in particular. Hooke went to see it
    on 2 June 1676 and was not impressed – 'Damned Doggs,' he
    wrote in his diary, '*Vindica me Deus* [Avenge me, God!]. People
    almost pointed' (Robinson and Adams, 1935, p. 235). There were

other Restoration plays that also took a potshot at science, but which left no mark on history, such as Thomas St Serfe's *Tarugo's Wiles: Or, the Coffee-House*, which had a nasty anti-Semitic flavour in certain passages (e.g. St Serfe, 1668, pp. 24–5).

33 For a general discussion of the history and importance of scientific societies, see Pyenson and Sheets-Pyenson (1999).

34 Shapin and Schaffer (1985).

35 Brown (1934); Harth (1983); Lux (1989).

36 Huygens (1974b), p. 325. See also McKeon (1965) and McClaughlin (1974, 1975).

37 Brown (1934); Hirschfield (1981).

38 See the brief autobiographical document in *Bibliotheca Thevenotiana* (Anonymous, 1694) and the obituary in *Journal des Sçavans*, October 1692. Virtually the only other sources relating to Thévenot are various contemporaries, in particular Steno, Swammerdam, Huygens, Chapelain and Leibniz. For the best summary of Thévenot's scientific life, see his entry in the *Dictionary of Scientific Biography*. For the descriptions of Thévenot see Chapelain to Bernier (13 November 1661), in Tamizey de Larroque (1883), vol. 2, p. 170; Huygens to Huygens (21 July 1661), in Huygens (1974a), p. 301. The Leibniz quote is from an unpublished manuscript by Nicolas Dew, 'Collecting travels in late seventeenth-century Paris'. After Thévenot's death, the English physician Martin Lister visited Paris and saw Thévenot's library. Lister wrote: 'This man was, as it were, the Founder of the *Académie des Sciences*, and was in his own nature very Liberal and gave Pensions to many Scholars.' Lister (1699), p. 104. Melchisedec is often confused with his nephew, Jean de Thévenot (1633–67), another orientalist who travelled widely and supposedly introduced coffee to France (De Crousaz-Crétet, 1922, p. 188), who wrote about his voyages and was eventually killed on the coast of Coromandel, eastern India, whilst on a commercial mission for the King (*La Gazette d'Amsterdam*, 20 September 1668). There is no known portrait of Melchisedec, although a picture of his nephew Jean, from the frontispiece to De Thévenot (1665), has been taken for him by Lindeboom (1975b).

39 *Relations de divers voyages curieux*, published in 1663, 1664, 1666 and 1672.

40 In a letter dated 21 July 1661 Christiaan Huygens describes Thévenot's address in Paris: '*Il demeure dans la rüe de Touraine qui est*

vers l'hotel de Guise, et c'est la première porte cochere a main droite' (Huygens, 1974, p. 301). The rue de Touraine au Marais is now the rue de Saintonge, in the 3rd *arrondissement*. Thévenot's house is apparently still there. In 1677 Swammerdam addressed letters to him at the rue des Vieilles Etuves (now under the esplanade of the Centre Pompidou) and there is evidence that in 1675 he was living on the rue de la Tannerie, near the Hôtel de Ville (Schiller and Théodoridès, 1968). Issy-les-Moulineaux is now a built-up suburb which flows seamlessly into Paris. The unmarked site of Thévenot's country house is now the garden of an old people's home.

41 Petit to Huygens (17 October 1664), in Huygens (1977), p. 124.

42 Huygens to Thévenot (27 November 1664), in Huygens (1977), p. 152.

43 '*Etant revenue à Paris, j'attirai chez moi une compagnie de personnes connuës pour très-habiles, entre lesquels Messieurs Frenicle & Stenon étoient logez chez moi. J'entretenois dans une maison jointe à la mienne une autre personne pour les expériences de Chymie; mais la dépense de ces expériences, de ces observations & de ces anatomies excedant de beaucoup mon revenu, après l'avoir soûtenu deux ans durant je proposai à M. Colbert de donner une forme plus durable à cette assemblée en la faisant agréer du Roi.*' Anonymous (1694), n.p. Chapelain to Steno (27 May 1667): '*Mr Thévenot s'est opiniastré, depuis dix-huit mois, à ne prendre point de maison à Paris pour philosopher et spéculer, dit-il, avec plus de liberté à la campagne*' (Tamizey de Larroque, 1883, vol. 2).

44 Thévenot to Huygens (January 1662), in Huygens (1974b), pp. 18–19. See also Nicholas Dew's unpublished manuscript 'Collecting travels in late seventeenth-century Paris'.

45 For the 'pulsation of air' see Oldenburg to Huygens (29 March 1662), in Hall and Hall (1965), p. 447. For the proposal to use the bee cell as a universal measure, see Thévenot to Oldenburg (29 October 1671), in Hall and Hall (1971), p. 312. Swammerdam discussed this proposal and pointed out that cell size was not constant (Swammerdam, 1758, pt 1, p. 164), but nevertheless used it to measure the length of the digestive tract of the gadfly larva (it was five cells long) (Swammerdam, 1758, pt 2, p. 48).

46 Dibon (1963).

47 '*Nous continuons tousjours a faire quelque chose chez Monsieur Thevenot principalement sur lanatomie a loccasion de Monsieur Stenonius qui est*

icy.' Petit to Huygens (28 November 1662), in Huygens (1974b), p. 270.

48 Swammerdam may have made the long journey to Saumur for spiritual reasons – he stayed with Tanneguy Le Febvre (1615–72), Professor of Greek and a well-known Protestant theologian, who had connections with Leiden University; see Lindeboom (1975a). One of Swammerdam's closest companions in Saumur was Isaac d'Huisseau, an amateur naturalist who later corresponded with Steno, who was also a fierce advocate of the unity of all Christians on a relatively limited, scriptural basis – this corresponded to Swammerdam's views a decade later, when a religious crisis led him to abandon science for some time. See Bourchenin (1884), p. 39 and Bourchenin (1887).

49 Nordström (1954–5), p. 35.

50 Cobb (2002a).

51 Nordström (1954–5), pp. 33–4.

52 Schiller and Théodoridès (1968). See, for example, the report of Steno's book on muscles in *Journal des Sçavans*, 23 March 1665.

53 Steno (1965).

54 '*Nous avons pris l'occasion du froid des mois passés et nous nous sommes apliqués a faire des anatomies et a examiner la Generation des animaux.*' Thévenot to Huygens (24 April 1665), in Huygens (1977), p. 343. This is the correct transcription.

55 In the dedication to Thévenot in his doctoral thesis *De Respiratione*, Swammerdam wrote, 'Never shall I forget the blessed refuge which I and our common friend Dr Steno found at your farm at Issy, where you obtained for us tranquillity for studies and everything we required so that we could investigate the growth of the egg, that marvel of nature' (Swammerdam, 1667, n.p.).

56 Stenonis (1675).

57 Steno (1950).

58 Descartes, *De la formation de l'animal*, quoted in Roger (1997), p. 115. The developmental biologist Lewis Wolpert has discussed the ultra-determinist modern equivalent of Descartes' idea (without, however, referring to Descartes): 'Is the egg computable? That is, given a total description of a fertilised egg – the total DNA sequence and the location of all proteins and RNAs – could one predict how the embryo would develop?' (Wolpert, 1995, p. 62). The short answer is 'No' – the embryo has to interact with the environment, the proteins produced by DNA also interact with

each other in both time and space, while some genes control the activity of other genes.

59 Christopher Wren also visited Paris at this time, and met Thévenot. Tinniswood (2001).

60 These details are given by De Graaf in the dedication to *Tractatus de Vivorum Organis Generationi Inservientibus* (De Graaf, 1668). For an annotated translation into English see *Journal of Reproduction and Fertility*, Supplement No. 17 (1972).

61 Grmek (1990), p. 244.

62 Moore (2003).

CHAPTER 3: INSECTS IN ITALY (pp. 63–93)

1 Redi (1909), p. 64.

2 'But the more curious one will say, That the *Scorpion* came from without, to the sweet smell and food of the *Herbe*: but that doubt is prevented. For truly, the two bricks being mutually beaten together, did suitably touch each other, so that they hindered the entrance of the *Scorpion*, as well by their co-touching plainness, as by their weight' (Van Helmont, 1662, pp. 113–14).

3 'Mr HOSKYNS related an experiment of the production of bees out of dead bullocks, described by Mr HARTLIB out of an extract of Mr CARY, in his *Commonwealth of Bees*. This was directed to be tried in a warm place, but in the shade' (Birch, 1756–7, vol. 1, p. 270). Nothing further was heard of this experiment.

4 For a summary of Greek and Roman views of spontaneous generation, see McCartney (1920).

5 Boyle (2000), p. 287.

6 Principe (1998).

7 Pagel (1958), p. 91.

8 See, for example, the contribution by Mr Henshaw to the Royal Society on 7 December 1664 (Birch, 1756–7, vol. 1, p. 501).

9 Birch (1756–7), vol. 1, p. 41.

10 At a meeting on 13 November 1661 (Birch, 1756–7, vol. 1, p. 53).

11 At a meeting on 9 March 1664 (Birch, 1756–7, vol. 1, p. 393).

12 This and subsequent quotes from Birch (1756–7), vol. 1, pp. 22, 36, 266, 112, 117, 118, 212, 238 and 448.

13 Ibid., vol. 2, pp. 48–9.

14 Technically, the journal was the property of the Royal Society's Secretary, Henry Oldenburg, but it was published under the authority of the Royal Society.

15 People obviously talked about these studies, however. Nearly thirty years later, John Ray remembered John Wilkins FRS telling him about this work. After describing Redi's experiments, Ray wrote: 'The same Experiment I remember Doctor *Wilkins* late Bishop of *Chester* told me had been made by some of the *Royal Society*.' Ray (1691), p. 221.

16 The background on Florence, the Medici and their attitude to science is taken from Cochrane (1973); Hibbert (1974); Litchfield (1986); Hale (2001); Strathern (2005).

17 Hibbert (1974) presents the traditional account of Lorenzo and Leonardo. Nicholl (2004) has a more critical view.

18 Bergin (2001), pp. 37–41.

19 Middleton (1971).

20 Boschiero (2002, 2003).

21 Cochrane (1973), pp. 231–313, contains a major discussion of Magalotti's life and work, including his activity with the Accademia del Cimento. See also Middleton (1971).

22 The *Saggi* are translated in Middleton (1971).

23 Middleton (1971), p. 78.

24 Tribby (1991); Biagioli (1992, 1995); Findlen (1993).

25 There is no English biography of Redi. There is an Italian website entirely devoted to Redi's life and work (www.francescoredi.it) which includes downloadable versions of all his major works. There is a discussion of the full range of Redi's work in Bernardi and Guerrini (1999).

26 Paula Findlen light-heartedly points out that 'most people can't pronounce' Kircher's name (Findlen, 2004, p. xi). This is her preferred version (Findlen, personal communication).

27 This is the subtitle of a collection of essays on Kircher (Findlen, 2004). For much of the last 250 years, Kircher has been a favourite with Rosicrucian sects, a bizarre figure who devised a 'cat organ' (you pressed the keys and they hit the cats' tails, which would make them squeal), possessed a stuffed basilisk and allegedly invented the magic lantern; a character who was later found lurking in the novels of Umberto Eco. At the turn of the twenty-first century, cultural historians began to reappraise Kircher and his

work. As Paula Findlen wryly put it, in a few years he metamorphosed from 'that strange Jesuit – the man who got everything wrong' – into 'the coolest guy ever' (Findlen, 2004, pp. ix–xi). Quotes are taken from Findlen (2004), pp. 22, 7, 6.

28 Oldenburg to Boyle (21 November 1665), in Hall and Hall (1966a), p. 615.

29 Oldenburg to Boyle (18 September 1665), in Hall and Hall (1966a). p. 512.

30 See Stephen Jay Gould's sensible discussion of Kircher's palaeontology (Gould, 2004).

31 Kircher, *Mundus Subterraneus*, XII, chap. 4, exp. 1. Translation given in Gottdenker (1979), p. 576. It is unclear what kind of 'microscope' Kircher was using; it probably had very low magnification.

32 Rowland (2004), p. 199.

33 Redi (1664). An English translation appears in Knoefel (1988).

34 This and subsequent quotes taken from Knoefel (1988), pp. 5, 7, 23, 3.

35 Quoted in Findlen (1993), p. 47.

36 Redi (1668). An occasionally inaccurate English translation of the 1688 edition, by Mab Bigelow, appears in Redi (1909). A better translation into French, by André Sempoux, together with annotations and translations of other writings by Redi, can be found in Redi (1970). On 14 July 1668 the Royal Society heard that Redi's book was nearly printed (Hall and Hall, 1967, p. 541). This implies it must have been finished several weeks earlier.

37 Dati was a philologist, and, like Redi, was a member of the Accademia del Crusca (The Bran Academy), a linguistic group charged with drawing up a Tuscan dictionary. The symbol of the Accademia del Crusca can be seen on the title page of Redi's 1668 book; he also refers to the dictionary being written by 'our Academy' (Redi, 1909, p. 33).

38 Ferdinando is mentioned only three times, in passing.

39 In 1671 Redi criticised Kircher for his belief in the curative power of stones allegedly found in viper heads. See Baldwin (1995).

40 The quotes that follow are from Redi (1909), pp. 23–37.

41 From his description, these tiny 'flies' would appear to have been small parasitoid wasps, which had parasitised the fly larvae.

42 The appearance of worms in cheese has a long mystical pedigree, although there is no evidence that Redi was aware of this. See

Ginzburg (1980).

43 This and subsequent quotes are from Redi (1909), pp. 35, 43, 64, 79, 62, 87 and 64.

44 Findlen (1993), p. 44.

45 Stone and Schönrogge (2003).

46 This and subsequent quotes are from Redi (1909), pp. 92–4.

47 Redi (1909), pp. 115–16.

48 Redi to Lanzoni (20 February 1693), in Bonnet (1783), p. 423n. This letter is also quoted in Gottdenker (1979), p. 579.

49 Redi to Steno (4 February 1667), in Redi (1970), p. 102.

50 Bonnet (1783), p. 422. Bonnet goes on, more perceptively, to suggest that 'When a happy Genius raises himself above his century, he always retains something of the century which preceded him, and the century in which he lives' (Bonnet, 1783, p. 423).

51 Quoted in Findlen (1993), p. 45.

52 Colbert, the French Finance Minister and the man who set up the Académie des Sciences, appears to have received a copy in May 1669. Everyone else had to wait. Oldenburg to Finch (4 September 1668), in Hall and Hall (1968), pp. 35–6; Justel to Oldenburg (23 February 1669), in Hall and Hall (1968), pp. 401–2; Oldenburg to Huygens (31 May 1669), in Hall and Hall (1968), p. 581; Paisen to Oldenburg (27 November 1669), in Hall and Hall (1969), p. 341; Oldenburg to Huygens (5 July 1669), in Hall and Hall (1969), p. 94. See also *Philosophical Transactions*, 1670, 5: 1175. Magalotti brought along some copies when he visited London in 1669 with Prince Cosimo. Justel to Oldenburg (9 December 1668), in Hall and Hall (1968), pp. 226–9.

53 Huygens to Oldenburg (16 June 1669), in Hall and Hall (1969), p. 46.

54 Ray (1671), p. 2220. This led to a letter from Martin Lister, who was also doubtful about Redi's idea (Lister, 1671).

55 Quotes in the rest of this paragraph from Gottdenker (1979), p. 580. This also suggests that, for Kircher at least, most of Redi's experiments were not carried out in the presence of the Tuscan court.

56 Birch (1756–7), vol. 3, p. 420. Hooke certainly possessed a copy of Redi's book (see the facsimile of the catalogue of his library, in Rostenberg, 1989, p. 181), which he received on 19 February 1673 (see his diary entry for that day, in Robinson and Adams, 1935, p. 29). That does not mean, of course, that he either read or

fully took on board its findings.

57 For contrasting accounts of this final phase in the history of spontaneous generation, see Farley (1977); Strick (2000); Harris (2001).

CHAPTER 4: THE TESTICLES OF WOMEN (pp. 94–124)

1 Steno (1668), p. 48. A facsimile, translation of and introduction to the section on muscles can be found in *Transactions of the American Philosophical Society* (1994), 84: 1–249.

2 Letter from Steno to Viviani, quoted in Cioni (1962), p. 56. Shortly after Steno's arrival in Florence, Viviani had to intervene to ensure that Steno was given more appropriate lodgings – in his initial accommodation he was continually pestered by 'impertinent women'. See Rome (1956), p. 524.

3 *Transactions of the American Philosophical Society* (1994), 84: 83.

4 Kardel (1994b).

5 It also contrasted with Swammerdam's empirical and experimental approach to the problem of muscle and nerve activity, carried out at around the same time, but which had a greater immediate impact. See Cobb (2002a).

6 A translation of *Canis Carchariae Dissectum Caput* can be found in Scherz (1969).

7 Scherz (1969), p. 73.

8 Ibid. (1969), p. 113–15.

9 Gould (1987), Cutler (2003).

10 Swammerdam (1672), p. 50.

11 Steno (1668), p. 117.

12 Anonymous (1667b).

13 Ibid. (1679b).

14 Barles (1674), n.p. (Epitre).

15 Stenonis (1675).

16 Bartholini (1680).

17 Historians of science have become increasingly interested in the meaning and impact of scientific illustrations. For a summary and a guide to further reading, see Cobb (2002b).

18 Dr Gunther von Hagen's modern 'plastination' technique is a direct descendant of these methods.

19 Cole (1921).

20 For an account of Ruysch's life and work, see Luyendijk–Elshout (1970).

21 Swammerdam (1672), p. 37.

22 All details from Swammerdam (1672), pp. 46–53.

23 Swammerdam (1672), p. 50.

24 All details here are taken from Cochrane (1973), Hibbert (1974).

25 Hoogewerff (1919), pp. 99–103; Swammerdam (1672), p. 50.

26 Swammerdam (1672), p. 51.

27 Details from Kardel (1994a).

28 Stenonis (1675).

29 Short (1977), p. 16.

30 De Graaf's *prodromus* is reproduced at the beginning of De Graaf (1673) and in De Graaf (1699), pp. 192–5.

31 Van Horne (1668). Van Horne's *prodromus* is reproduced, with comments, in Swammerdam (1672), in De Graaf (1673), and is included under the title '*Anatomicum veterarum exercitatissimum, suarum circa partes generationis in utroque sexu observationum prodromus*', in Van Horne (1674), pp. 116–26.

32 Swammerdam (1672), p. 15; Van Horne (1674), p. 124.

33 Anonymous (1668). At its meeting of 19 March 1668 the Royal Society decided that Ent 'should be desired to peruse these books, and to give his thoughts of them to the Society'. Birch (1756–7), vol. 2, p. 259. It seems most likely that Ent was the author of the review.

34 This and subsequent quotes from De Graaf (1668), n.p. (Preface). Translated in Jocelyn and Setchell (1972), p. 5.

35 Anonymous (1878), p. 29. No source is given for Sylvius's letter.

36 Swammerdam to Thévenot (24 September 1665), in Lindeboom (1975b), pp. 39–41. For a discussion of this, see Nordström (1954–5).

37 Colepresse to Oldenburg (14 December 1668), in Hall and Hall (1968), p. 248.

38 Ibid. (16 August 1669), in Hall and Hall (1969), p. 194; Cook (1994), p. 70. Sylvius wrote a book about the epidemic. For a review, see Anonymous (1671c), pp. 2212–13.

39 De Graaf (1668), n.p. (Preface). Translated in Jocelyn and Setchell (1972), p. 7.

40 De Graaf (1668). A full translation, together with an introduction and notes, can be found in Jocelyn and Setchell (1972). A translation of De Graaf's *Tractatus de Clysteribus* ('On enemas'), together

with a brief introduction, can be found in Brockbank and Corbett (1954). A French translation, together with a biographical account, can be found in De Graaf (1878).

41 De Graaf (1668), n.p. Translation in Jocelyn and Setchell (1972), p. 9. Although this might sound very modern, the translators suggest that it 'has the smell of St Thomas and medieval scolasticism' (Jocelyn and Setchell, 1972, p. 65, note 10).

42 De Graaf (1668), p. 141. Translation in Jocelyn and Setchell (1972), p. 49.

43 Ibid. (1668), p. 118. Translation in Jocelyn and Setchell (1972), p. 42.

44 Ibid. (1668), p. 124. Translation in Jocelyn and Setchell (1972), p. 44.

45 Ibid. (1668), p. 160. Translation in Jocelyn and Setchell (1972), p. 54.

46 Ibid. (1668), p. 57. Translation in Jocelyn and Setchell (1972), p. 25.

47 Clarck (sic) (1668). The original is in Latin; for a reproduction and an English translation see Clarke to Oldenburg (April/May 1668), in Hall and Hall (1967), pp. 350–69.

48 Pepys' diary, 11 May 1660.

49 Pepys' diary, 28 April 1662 (Latham, 1985, p. 192). Pepys' initial candidacy failed. He was not admitted until February 1665, when he was successfully proposed by Thomas Povey.

50 Denis had recently been put on trial for murder, after one of his transfusion operations went wrong. See Moore (2003).

51 Clarke to Oldenburg (April/May 1668), in Hall and Hall (1967), p. 366.

52 Ibid. (April/May, 1668), p. 367.

53 The event was seen as so catastrophic that Pepys feared an imminent attack on London and immediately arranged to send his wife and father to the country 'with about £1300 in gold'. Over the next few days he reported rumours 'that we are bought and sold, that we are betrayed by the papists and others about the King'. In this atmosphere of fear and paranoia, the Secretary of State, Lord Arlington (one of the 'A's in the 'CABAL' of five ministers that ran the country), ordered a round-up of suspects and scapegoats – mainly foreigners and Catholics. Oldenburg got caught in the net. On 25 June, Pepys recorded in his diary that his colleague had been 'put into the Tower for writing news to a Virtuoso in France

with whom he constantly corresponds in philosophical matters; which makes it very unsafe at this time to write, or almost do anything'. Oldenburg was not charged; indeed, it was several weeks before he learnt why he was in prison at all. It turned out that he was suspected of 'dangerous desseins and practices ... inferred from some letters and discourses'. He had not been protected by the Society's Royal Charter of 1662, which allowed him to engage in wide-ranging overseas correspondence, nor by the transparent subterfuge of having letters addressed to 'Mr Grubendol, London' (addresses were much simpler in those days). Luckily for Oldenburg, for the Royal Society, and indeed for the history of science, he was released without charge (or apology) after six weeks and wrote a suitably grovelling letter to the King who had overseen his arrest. Who had informed on him, and which of his letters had been intercepted, was never revealed. For more details, see Hall (2002).

54 De Graaf to Oldenburg (25 September 1668), in Hall and Hall (1968), pp. 67–71.

55 Birch (1756–7), vol. 2, p. 317.

56 Ibid. (1756–7), vol. 2, p. 321.

57 Ibid. (1756–7), vol. 2, p. 333. Quotes from Clarke's reply to De Graaf in this paragraph are from Clarke to Oldenburg (20 December 1668), in Hall and Hall (1968), pp. 268–72.

58 De Graaf to Oldenburg (12 February 1669), in Hall and Hall (1968), p. 400.

59 Oldenburg to De Graaf (8 May 1669), in Hall and Hall (1968), p. 530.

60 Colepresse to Henry Oldenburg (16 August 1669), in Hall and Hall (1969), p. 193. Samuel Colepresse was apparently one of the thousands of victims of the disease that also carried off Van Horne.

61 De Graaf to Oldenburg (15 July 1669), in Hall and Hall (1969), p. 122.

62 Birch (1756–7), vol. 2, p. 397.

63 Anonymous (1669) and Figure II in the same issue.

64 Clarke to Oldenburg (20 December 1669), in Hall and Hall (1969), p. 389. De Graaf replied in June 1670, but managed to focus on the disputed anatomical points. Clarke had shifted the issue away from the structure of the testicle to the fine detail of the male vas deferens – a vessel that transports sperm. De Graaf to Oldenburg (19 June 1670), in Hall and Hall (1970), pp. 35–40.

65 In his final letter on the affair Clarke complained to Oldenburg: 'When you communicated to me his letter to you of 5 October and urged me that I should not only uphold the justice and fairness of the Royal Society but should also explain myself better concerning some matters in that letter of mine published by you, I submitted to your incitements, unluckily enough, since you like me were misled when we believed that Mr. De Graaf had rebuked the Royal Society; and I to my own misfortune chose words so badly or arranged them so ill that in explaining myself I stirred up greater or at any rate further quarrels not only in that but in another later letter.' Clarke to Oldenburg (20 December 1669), in Hall and Hall (1969), p. 386. For a discussion of Oldenburg's role at the centre of his web of correspondence, see Lux and Cook (1998). I have borrowed the concept of a 'network' from this article.

CHAPTER 5: THE RULES AND THEOREMS OF GENERATION (pp. 125–54)

1   Needham (1959), p. 34.
2   Van Speybroeck et al. (2002), p. 9.
3   Needham (1959), p. 47.
4   Ray (1691), p. 217.
5   Petersson (1956).
6   This and subsequent quotes are taken from Digby (1644), pp. 203, 204, 222, 219, 223.
7   This and subsequent quotes are taken from Highmore (1651), pp. 84, 91, 82.
8   Adelmann (1942).
9   This and subsequent quotes are taken from Harvey (1981), pp. 244, 240, 136, 206, 203–4, 202.
10 For a brief summary of Mouffet/Moffett/Mouset/Muffett's life, see his entry in the *Oxford Dictionary of National Biography*; Houliston (1989a); Dawmarn (2003). It is sometimes suggested that Mouffet's daughter Patience was the child in the children's nursery rhyme 'Little Miss Muffett'. Sadly, it seems unlikely – the first appearance of the nursery rhyme in the 'Miss Muffett' version seems to have been in the late nineteenth century.
11 Moffett (1634).

12 Muffet (1658), pp. 890, 897.
13 Many of the views described in the *Theatrum* were the work of Thomas Penny, an Anglican priest with a fascination for insects and plants, who in 1565 had found himself in trouble with his religious superiors because of his radical puritan views. Hoping to make a new life for himself, he travelled to Zurich where he met Conrad Gessner (1516–65), the great Swiss bibliographer and zoologist. The two men struck up a close friendship, but Gessner died shortly afterwards, leaving some of his manuscripts to Penny, who brought them back with him when he eventually returned to England. When Penny died in 1589, he left all his papers, including Gessner's writings, to his friend and physician, Mouffet. Mouffet put these studies, together with material by Edward Wotton (1492–1555) and some of his own writings, into the *Theatrum*, giving the three other men equal credit on the title page. To complete the complex story, Mouffet died in 1604, with the book, originally dedicated to Elizabeth I, still unpublished. While the reign of Elizabeth's successor, James I, came and went, the book 'lay for a long time in obscurity under the custody of friends of the Author departed' before finally coming into the hands of the alchemist and court physician Sir Theodore Mayerne, who saw it through to publication in 1634, dedicated to Charles I.
14 Reprinted in 1989 (Houliston, 1989b). All details are taken from the introduction to the facsimile edition, by Victor Houliston. Over the years Mouffet's poem has caused some excitement amongst Shakespeare scholars, for it contains the story of Pyramus and Thisbe, which also appears in *A Midsummer Night's Dream*. However, it is now generally accepted that there is no connection between the two works. Mouffet's poem was part of a brief English tradition of long, semi-serious poems on everyday subjects – around the same time two similar efforts appeared: *Lenten Stuffe* by Thomas Nashe, which was about the red herring, and Sir John Harrington's *The Metamorphosis of Ajax*, which sang the praises of the water closet.
15 Duchess of Newcastle (1671), p. 127.
16 This section of his book was reprinted as Libavius (1661).
17 Cherry (1989).
18 Just up the coast in The Hague, similarly precise work was produced by Jacques de Gheyn II, whose ink drawings took the

long-established tradition of miniature painting and fused it with the use of the magnifying glass to provide vivid close-up portraits of flowers and insects. For a sumptuous account of De Gheyn's work, see Swan (2005). For discussions of the interaction of art and microscopy, see Alpers (1983), Ruestow (1995) and Cobb (2002b).

19 Goedaert's way of drawing his figures at separate times has been pointed out by Brian W. Ogilvie in his unpublished manuscript 'Description and persuasion in seventeenth-century entomological illustrations'.

20 Goedaert (1662–8). See Ruestow (1995), pp. 53–9.

21 Freedberg (1991).

22 This and subsequent quotes are taken from Goedaert (1682), pp. 4, 10–11, 65, 49.

23 Boerhaave (1744), vol. 5, pt 2, pp. 495–6. Translation adapted from Adelmann (1965), vol. 2, p. 877, note 1.

24 Brown (1934), pp. 280–81.

25 Swammerdam (1685), p. 25; Swammerdam (1758), pt 1, p. 9. Swammerdam's *Historia Insectorum Generalis* was translated into French after his death (Swammerdam, 1685). Page references are to this edition. Much of the *Historia Insectorum Generalis* was included in Swammerdam's posthumous masterpiece, *The Book of Nature* (English translation, Swammerdam, 1758). In general, I have used the eighteenth-century English translation, and given the appropriate page references to both French and English editions. Occasionally, I have preferred to translate directly from the French, consulting the original Dutch edition.

26 Goedaert (1682), p. 4.

27 Brown (1934), pp. 280–81.

28 Anonymous (1667a), p. 535.

29 Swammerdam (1669); Oldenburg to Boyle (17 September 1667), in Hall and Hall (1966b), p. 476.

30 Swammerdam (1685) p. 47; Swammerdam (1758), pt 1, p. 15.

31 Ibid. (1685), p. 10; ibid. (1758), pt 1, p. 4.

32 Ibid. (1685), p. 131; ibid. (1758), pt 2, p. 30.

33 Wilson and Doner (1937).

34 Swammerdam (1685), p. 140; Swammerdam (1758), pt 2, p. 35.

35 Ibid. (1685), p. 159.

36 Ibid. (1758), pt 2, p. 37.

37 Ibid. (1685), p. 53; ibid. (1758), pt 1, p. 18.

38 Ibid. (1685), p. 37; ibid. (1758), pt 1, p. 12. Referring to the section of Harvey's *De Generatione* which dealt with insects, and which was destroyed during the English Civil War, Swammerdam assumed that Harvey's 'book on insects, which we are so sadly lacking, and for which we sigh with such ardour, contained curious experiments rather than certain truths, and that he did not explain the nature of the changes that occur in these small animals' (Swammerdam, 1685, p. 37). Swammerdam cut this phrase when preparing material from *Historia Insectorum Generalis* for what became *The Book of Nature*.

39 Ibid. (1685), p. 39; ibid. (1758), pt 1, p. 13.

40 Thanks to Casper Breuker and Melanie Gibbs for this translation. The Dutch original reads: '*deese swelling, uytbotting, uytpuyling, knopping, ende als spening van nieuwe leedmaaten, die allenkxkens door een toefetting van deelen, ende geenfins door vervorming, sijn aangegroeit ende gebooren*' (Swammerdam, 1737–8, vol. 1, p. 29). Swammerdam's translators found this sentence, which was also used in *The Book of Nature*, difficult to translate. The 1758 English translation of *The Book of Nature* has: 'In this inflation, (shooting out, budding, or vegetation; and, as it were, changing of the nutriment of the new limbs, which have gradually grown, or have been produced by an epigenesis, or accretion of the parts, and not at all by a metamorphosis)' (Swammerdam, 1758, pt 1, p. 13). The French translation of this passage from *Historia Insectorum Generalis* is perfunctory: '*l'unique fondement de tous les changemens, qui arrivent aux insects, ne consiste pas dans cette prétenduê transformation, mais seulement dans les bourgeons ou dans les boutons qui poussent ces nouveaux membres en croissant insensiblement les uns apres les autres*' (Swammerdam, 1685, pp. 39–40).

41 Swammerdam (1685), pp. 25–6; Swammerdam (1758), pt 1, p. 9.

42 Ibid. (1685) p. 13; ibid. (1758), pt 1, p. 5.

43 Ibid. (1685) p. 20; ibid. (1758), pt 1, p. 7.

44 Ibid. (1685), p. 20; ibid. (1758), pt 1, p. 7.

45 Ibid (1685), p. 38; ibid (1758), pt 1, p. 13.

46 This and subsequent quote from Swammerdam (1758), pt 2, pp. 112–13.

47 The 'grains' may have been denatured proteins, produced by the boiling and infusing he had used to harden and colour the embryos, or perhaps zones of pigmentation. Swammerdam apparently later confused them with cells – he wrote that the grains

'went likewise to compose the skin and internal parts of the Frog. A circumstance that could not fail of surprising me greatly, and the more so, as they were considerable and distinct enough to be seen with a common microscope' (Swammerdam, 1758, pt 2, p. 115). This indicates how difficult it can be to interpret what was observed in the past.

48   Swammerdam (1685), p. 47; ibid. (1758), pt 1, p. 16.
49   For discussions of Swammerdam's views of generation from a philosophical point of view, see Ruestow (1985, 1995) and Fouke (1985).
50   Swammerdam (1685), p. 137; ibid (1758), pt 2, p. 33.
51   Ibid. (1685), p. 163.
52   Theodore Mayerne, 'The Epistle Dedicatory', in Topsel (1658), n.p. Swammerdam's riposte can be found in Swammerdam (1685), p. 26; Swammerdam (1758), pt 1, p. 9.
53   For details of the relations between Swammerdam and Malpighi, see Cobb (2002b).
54   Anonymous (1670).
55   Malpighi (1669).
56   Ibid. (1669), p. 58.
57   Malpighi acknowledged that it would have been 'the greatest of chances if the experiment had worked as hoped' (Malpighi, 1669, pp. 83–4).

CHAPTER 6: LIFE AND DEATH (pp. 155–87)

1   Normally the Council met in Arundel House, but Brouncker 'declared that for this time he had caused the Council to be summoned in this place for his particular conveniency, his present occasions not haveing permitted him to go afarr off', Royal Society Archives, Minutes of Council, vol. 1, 1668–81, p. 216. (See also Birch, 1756–7, vol. 3, p. 93.) The 'present occasions' were presumably related to the third Anglo-Dutch war; the meeting would therefore have taken place in Lord Brouncker's office, in the Navy Board's new home in Derby House. In January 1673 the Board's building in Seething Lane in the City had been destroyed by a fire that began in Brouncker's apartment and 'which in six hours time Laid in ashes the said office with Severall of the houses about it'. Quoted in Tomalin (2002), p. 297. Rumour had it that

Brouncker's mistress, Mrs Williams, was responsible. Given that Pepys' house was also destroyed, and he had been able to save only his precious books (including his secret diary), his dislike of his superior's live-in companion must have increased substantially. After some time in a makeshift home in Mark Street, the Board had recently been moved again, this time to Derby House, just down the road. For the problem of dating in this period, see chapter 1, note 28, above. The details about the weather on this day from John Gadbury's diary of the weather in London (Gadbury, 1691). This reference was kindly provided by Ian MacGregor, Archive Information Officer at the Meteorological Office in Bracknell, Berkshire, U.K.

2 One of Brouncker's fancy buckles is on display at the Royal Society in London. Other material in this paragraph is from Hooke's diary (Robinson and Adams, 1935).

3 Royal Society Archives, Letter Book O, vi, p. 241. The scribe has incorrectly written down the title of Swammerdam's book. It should be *Miraculum Naturae* ('Miracle of nature') not *Miraculum Mundi* ('Miracle of the world').

4 Kerckring (1671). Kerckring's pamphlet was bound together with copies of his previous work on the skeletal development of the human foetus, and can generally be found at the end of Kerckring (1670).

5 Anonymous (1672a).

6 Anonymous [Jean Gallois] (1672).

7 *Journal des Sçavans*, 21 March 1672, p. 66. 'Gaulois' (sic) was identified as the author of these criticisms in Anonymous (1672a), p. 4024.

8 Anonymous (1672a).

9 Ibid. (1672a).

10 Swammerdam (1672), p. 25.

11 See, for example, the letter from d'Andrea to Malpighi (25 February 1673) in Adelmann (1975), vol. 2, pp. 635–43.

12 Kerckring was able to date his foetus with such precision because the husband of the dead woman stated that her period had finished three days before her death, and they had sex in the intervening time (Anonymous, 1672a, p. 4021). Whatever he dissected, it was not a three-day-old embryo. At this stage the embryo has only eight cells and is smaller than a pinhead.

13 This allegation, made by Charles Drelincourt in 1684, is reported

in Roger (1997), p. 210. See Pineau (1650), pp. 112–20. Pineau's brief document was written in Paris in June 1579.

14 Anonymous (1672a), p. 4025.

15 Ibid. (1671a), p. 2218.

16 Ibid. (1671b).

17 This and subsequent quotes from exchanges between Swammerdam and De Graaf are given in De Graaf (1673). I have used the French translation which appeared in De Graaf (1699), pp. 202–3. For details of the contacts between Steno and De Graaf, see Adelmann (1965), p. 347.

18 De Graaf (1671). The final page of the original text (p. 208) carries a footer – indicating that the next page was intended to be the Index. In the final version, page 209 is the beginning of De Graaf's letter to Schact.

19 Ricci (1950).

20 De Graaf (1671), p. 212.

21 Swammerdam (n.d.; 1672?). The only copy I have been able to locate is in the British Library, BL 748g 10(1).

22 Garmann (1672).

23 De Graaf (1672). A facsimile was published in 1965, as number XIII in the series *Dutch Classics on the History of Science* (Nieuwkoop: B De Graaf), together with an Introduction by J. A. van Dongen. Jocelyn and Setchell (1972) contains an annotated translation. Apart from the Introduction by Van Dongen and the notes by Jocelyn and Setchell cited above, there are many works which discuss De Graaf's book, including Catchpole (1940); Guyénot (1941); Lindeboom (1973); Setchell (1974); Short (1977); Laqueur (1990); Ankum, Houtzager and Bleker (1996); Sawin (1997); Blackledge (2003). For discussions of the immediate impact of De Graaf's work, see Adelmann (1965) and Roger (1997).

24 Hooke's diary entry for 16 August 1672, in Robinson and Adams (1935), p. 5.

25 De Graaf (1672), p. 77. Translated in Jocelyn and Setchell (1972), p. 106.

26 However, like Swammerdam, the way he presented the anatomy of the female genital organs was still influenced by the traditional view that the structures of the penis and of the vagina were related. For a general discussion of changes in the representation of the female anatomy at this time, see Laqueur (1990), who deals with the specific question of De Graaf's images on pp. 158–9.

27 De Graaf (1672), p. 81. Translated in Jocelyn and Setchell (1972), p. 107.

28 Ibid. (1672), p. 27. Translated in Jocelyn and Setchell (1972), p. 92.

29 Ibid. (1672), p. 180. Translated in Jocelyn and Setchell (1972), p. 133.

30 See, for example, Klein (1970–80).

31 De Graaf dissected rabbits 30 minutes, 6 hours, 27 hours, 48 hours, 52 hours and 3, 5, 6, 9, 10, 12, 14 and 29 days after mating.

32 De Graaf had provided an accurate description of the regressing empty follicle, which is now confusingly called the *corpus luteum* ('yellow body'). In most animals (including humans and rabbits) the empty follicle is ash-coloured or grey. It is yellow in the cow, in which Malpighi first named the structure.

33 De Graaf (1672), p. 313. Translated in Jocelyn and Setchell (1972), p. 166.

34 Ibid. (1672), p. 317. Translated in Jocelyn and Setchell (1972), p. 167.

35 Ibid. (1672), n.p. (Preface). Translated in Jocelyn and Setchell (1972), p. 82.

36 Ibid. (1672), p. 212. Translated in Jocelyn and Setchell (1972), p. 141.

37 Ibid. (1672), p. 237. Translated in Jocelyn and Setchell (1972), p. 148.

38 Ibid. (1672), n.p. (Preface). Translated in Jocelyn and Setchell (1972), p. 81.

39 Croxatto (2002).

40 De Graaf (1672), p. 327. Translated in Jocelyn and Setchell (1972), p. 170.

41 De Graaf (1672), p. 178. Translated in Jocelyn and Setchell (1972), p. 133.

42 See, for example, Garden to Leeuwenhoek (24 August 1693), in Van Leeuwenhoek (1976), p. 43.

43 De Graaf (1672), n.p. (Preface). Translated in Jocelyn and Setchell (1972), p. 81.

44 Ibid. (1672), p. 247. Translated in Jocelyn and Setchell (1972), p. 150.

45 When the oocyte is released it is accompanied by hundreds of cumulus cells from the follicles which gather round it and aid fertilisation. This might perhaps have been visible to the naked eye, and

could have been seen with a magnifying glass or a microscope, although De Graaf never mentioned using either instrument. It is clear that in his dissections of the rabbit four days after mating he saw a 'blastocyst' – the embryo prior to implantation in the uterus wall. Looking at eggs he found 'floating' in the uterus, he could see that 'in the confined centre of this limpid egg there is as it were another egg swimming about' (De Graaf ,1672, p. 263. Translated in Jocelyn and Setchell, 1972, p. 153). This is a good description of the early blastocyst, with the 'limpid egg' corresponding to the *zona pellucida* that surrounds the inner cell mass, which can be said to 'swim about' within it. Although we will never be certain of all that De Graaf saw, for the simple reason that his descriptions are often not precise enough, some of his observations are extremely telling. His observations on a number of species led him to conclude that around the time the embryo implanted into the uterus it was 'always spherical', but that it varied in size depending on the animal, being 'the size of a poppy seed, of a mustard grain, of a hemp seed, of a small pea and sometimes of a large chick pea. The matter seems to be about the same in all animals, fine and full of spirit and so limpid as to surpass entirely even the purest crystal' (De Graaf, 1672, pp. 262–3. Translated in Jocelyn and Setchell, 1972, p. 153). Equally impressively, De Graaf claimed that 'eggs just expelled from the ovaries are ten times smaller than others still adhering to the ovaries'. This is strikingly accurate – the oocyte and its surrounding cumulus cells take up around 10 per cent of the volume of the follicle in many mammals.

46 De Graaf's argument was technical rather than developmental. He argued: 'there is no such thing as a human foetus three or four days after coitus': fertilisation took place in the follicle shortly after copulation, with the fertilised egg ('the beginning of a nascent fetus') being released from the follicle subsequently. This took three days in rabbits and, he assumed, much longer in humans (De Graaf (1672), p. 328. Translated in Jocelyn and Setchell (1972), p. 170).

47 Archives of the Royal Society. A copy also exists in the British Library (Swammerdam, n.d.). The copy in the Royal Society archives is, as the label has it, 'mounted on woodden boards' (sic).

48 Birch (1756–7), vol. 3, p. 41.

49 De Graaf (1699), p. 212.

50 Robinson and Adams (1935), p. 252.

51 Swammerdam (1672), p. 22.

52 Ibid. (1672), n.p. (Dedication).

53 Ibid. (1672), p. 45.

54 Ibid. (1672). This copy, bearing the inscription of the person (Oldenburg?) who received it in London on 12 June 1672, is still in the Royal Society library.

55 This is the description of the uterus in the Royal Society's museum catalogue. Grew (1686), p. 8.

56 Oldenburg to Swammerdam (19 December 1672), in Hall and Hall (1973), p. 368.

57 Ibid.

58 Birch (1756–7), vol. 3, p. 312.

59 Swammerdam to Oldenburg (14 January 1673), in Hall and Hall (1973), p. 412. For details of what exactly Swammerdam's 'turpentine' might have been, see Cook (2002). With the help of the staff at the Royal Society, the British Museum and the Natural History Museum, I have attempted to track down the specimens, or at least discover what happened to them. In vain.

60 Jones (1996); Israel (1998).

61 De Graaf to Oldenburg (12 July 1673), in Hall and Hall (1973), p. 138.

62 This forms the dramatic opening scene of Alexandre Dumas' novel *The Black Tulip*.

63 These massacres were the subject of a famous series of horrifying engravings by Romeyn de Hooghe which were published in De Wiquefort (1673) and one of which is reproduced in Schama (1991), pp. 280–81.

64 Leoniceni (1673).

65 Swammerdam also identified Pechlin as the author in a confused letter written to Henry Oldenburg on 11 April 1673. Swammerdam to Oldenburg, in Hall and Hall (1973), p. 584.

66 Rey (1930), p. 79.

67 Oldenburg to Swammerdam (13 June 1672), in Hall and Hall (1973), p. 105.

68 This is a rough translation which captures the spirit of the insult, if not of the pun, contained in the original Latin: '*libellum tuum natibus haud naribus dignum*' – De Graaf (1673), p. 75. This sentence is often taken to refer to Swammerdam, but the context shows that he is referring to an unnamed person who was attacking Sylvius's memory. This was clearly Pechlin, not Swammerdam. The

offending sentence was not included in the French translation (De Graaf, 1699).

69 Meeting of 14 May 1673. Birch (1756–7), vol. 3, pp. 88–9.

70 This principle would soon be tested in the furious row which erupted in 1675 between Hooke and Huygens over who had first invented the balance-spring watch. This priority dispute has been studied in great detail, partly because of Hooke's ferocious defence of what he saw as his priority in the matter. See, for example, Jardine (1999).

71 Birch (1756–7), vol. 3, pp. 102–7. The final point is number 10, but there is no point 5.

72 The committee's report is dated 15 August 1673 (old style). De Graaf died on 17 August 1673 (new style). Because of the different calendars used in England and the Dutch Republic (see chapter 1, note 28), 15 August in London was 25 August in Delft; 17 August in Delft was 7 August in London.

73 Leeuwenhoek to Garden (19 March 1694) in Van Leeuwenhoek (1976), p. 59. A different version of this story is given by J. A. van Dongen in his introduction to the 1965 facsimile of De Graaf's *De Mulierum Organis Generationi Inservientibus* (Nieuwkoop: De Graaf): 'Albrecht von Haller says that there was a quarrel between de Graaf and Swammerdam at the home of Antonie van Leeuwenhoek, in consequence of which de Graaf died within twenty-four hours' (p. 24). Unfortunately, Van Dongen gives no source for this story. In his bibliography, he refers to two of Haller's works – *Bibliotheca Anatomica* and *Bibliotheca Medicinae Practicae* (1774 and 1779, respectively). These are merely summaries of medical books; I have consulted both works extensively and can find no mention of Leeuwenhoek at all, and no biographical information for De Graaf, beyond that contained in Haller's summary of *Partium Genitalium Defensio*. I have therefore been unable to verify what seems to be a very unlikely story. Closer to the events, the Dutch physician Herman Boerhaave, who later wrote a brief biography of Swammerdam and was close friends with Ruysch (who knew all the people involved), said that De Graaf was 'taken off by the plague' (Boerhaave (1719), p. 223). Kees Jansen (personal communication) has raised the possibility that De Graaf may have committed suicide, given his undoubtedly depressed state of mind at the end of his life. There is no evidence for this, and the fact that he was buried in consecrated ground argues strongly against the

possibility. It seems very unlikely that we will ever know how he died.

74 Journal Book Original, vol. V, p. 42, Royal Society Archives. A slightly amended version of this minute is given in Birch (1756–7), vol. 3, p. 102.

75 Swammerdam referred to his work with Van Horne, in a brief aside in his study of bees, but he made no claim to have been the first to discover that humans come from eggs. Swammerdam (1758), pt 1, p. 159.

76 Birch (1756–7), vol. 3, pp. 102–7.

77 Anonymous [Henry Oldenburg?] (1672). There was no review of De Graaf's *Partium Genitalium Defensio.*

78 De Diemerbroeck (1694), p. 156. The Latin original was published in 1674.

79 Portal (1770), p. 339.

80 Guyénot (1941), p. 252.

81 Swammerdam used the microscope extensively to study insects, but does not appear to have used it to investigate the human egg. Although De Graaf does not mention using a microscope, some of his descriptions of the blastocyst are so precise as to suggest he used some kind of instrument. The title page of his book shows *putti* using a primitive microscope, but there is no reason to take this as anything more than a symbol. Kerckring was positively critical of the use of the microscope in anatomy (Kerckring, 1670, pp. 177–9). For the only modern attempt to summarise the polemic between De Graaf and Swammerdam, which adopts the traditional view and comes down on De Graaf's side, see the long editorial note in Hall and Hall (1973), pp. 586–8, note 3.

82 Roger (1997), pp. 210–35.

83 De Baar (1994); Cobb (2000).

CHAPTER 7: MAN COMES NOT FROM AN EGG
(pp. 188–219)

1 De Graaf to Oldenburg (18 April 1673), in Hall and Hall (1973), p. 603.

2 From 1685 onwards Leeuwenhoek started calling himself 'van' Leeuwenhoek, which is how he is more widely known today. Since the 1930s, his life and discoveries have been more intensely

studied than those of any other member of the generation network. For summaries of his life and astonishing contribution to science, see Dobell (1932); Schierbeek (1959); Palm and Snelders (1982); Hall (1989); Boutibonnes (1994); Ruestow (1995); Wilson (1995); Fournier (1996). There is a nineteen-volume Dutch-English series of Leeuwenhoek's *Collected Letters*, which provides a rich resource for studying his work. In the 1980s, Brian Ford found some of Leeuwenhoek's samples in the archives of the Royal Society. This remarkable discovery, together with a discussion of the power of his microscopes, is described in Ford (1991).

3 In 1628, Johannes Faber, a member of the Accademia dei Lincei, wrote: 'A few days ago I looked through an optical tube of marvellous clarity, and was astonished by what I saw. It was made with great skill and craft by two Germans who brought it to my house and presented it to me. Since it was made for the observation of very small things, I decided to call it a *microscope*, by analogy with the telescope.' Quoted in Freedberg (2002), p. 183. Bignami (2000) mistakenly suggests it was Stelluti who coined the term.

4 Stelluti's detailed copper engravings were first printed in 1625 on a single sheet with limited circulation; a second edition was incongruously hidden away inside a book of translation of the Roman poet Persio (Stelluti (1630), pp. 51–4). For a full discussion of the Accademia dei Lincei's work with the microscope, including an explanation of why Stelluti published his illustrations in a book of poetry, see Freedberg (2002).

5 Hodierna (1644); Fontana (1646); Borel (1656).

6 I. W. [John Wilkins] (1648), p. 115.

7 Manuscript by Matthew Wren (Christopher's cousin), quoted in Jardine (2002), pp. 99–100.

8 Henry Powle offered his services to the Society; it was suggested he should speak to Christopher Wren about 'the lunar globe, and the pictures of small insects by the miscroscope (sic)'. Birch (1756–7), vol. 1, p. 33.

9 Minutes of meeting of 25 March and 1 April 1663, Birch (1756–7), vol. 1, pp. 213–15.

10 Hooke (1665). Over the subsequent centuries a number of facsimiles have been produced, including a recent CD-ROM. The entire text of *Micrographia*, including the stunning illustrations, is available free on-line at www.gutenberg.org. A few months earlier another member of the Royal Society, Henry Power, had published his 'New

Experiments Microscopical' in his book *Experimental Philosophy* (Power, 1664). Although this dealt with similar material to Hooke – studies of flies, bees, sparks and fleas – it was inferior in most respects, and above all, had only a few pictures, none of them approaching the impact of Hooke's. Power, who lived in Halifax (Yorkshire) had carried out most of his observations by August 1661, and demonstrated some of them to a meeting of the Royal Society in June 1663. Power also wrote a poem about the microscope (reproduced in Cowles, 1934). For a study of the contrasting approaches of Power and Hooke, see Knellwolf (2001). For a positive appreciation of Power, including the suggestion that, if his book had had illustrations, he would now be as well known as Hooke, see Hall (1966).

11 Pepys' diary, 21 January 1665 (Latham, 1985, p. 464).

12 Huygens to Moray (27 March 1665), in Huygens (1893), p. 282.

13 Hooke's work has been the subject of a great deal of study by historians of science and art, including Alpers (1983); Dennis (1989); Harwood (1989); Wilson (1995); Fournier (1996); Knellwolf (2001); Inwood (2002); Bennet et al. (2003); Jardine (2003); Chapman (2005); Turner (2005).

14 Huygens to Hudde (4 April 1665), in Huygens (1893), p. 304. Huygens must have known some very large cats – the fold-out flea and louse are each about half a metre long.

15 Ford (2000).

16 Hudde, the man who may have invented the single-lens microscope, complained to Christiaan Huygens that he found *Micrographia* frustrating because he could not read English; he asked Huygens to translate some sections of it for him. Hudde to Huygens (5 April 1665), in Huygens (1893), p. 309.

17 Pepys' diary, 13 August 1664 (Latham, 1985, p. 415). With his wife, Pepys 'found great difficulty before we could come to find the manner of seeing anything by this microscope' (entry for 14 August 1664).

18 Whitaker (2003).

19 Duchess of Newcastle [Margaret Cavendish] (1666), n.p. (Preface). For discussions of the Duchess's ideas, see Rogers (1996); Price (2002); Whitaker (2003).

20 Leeuwenhoek to Hooke (5 April 1680), in Van Leeuwenhoek (1948), p. 215.

21 For full details of Leeuwenhoek's time as the curator of the Vermeer estate, see Montias (1989), pp. 225–30; for extracts from

the original legal documents, see documents 370, 379 and 416. Leeuwenhoek's task as curator was not easy – as late as 1682 he was still administering the Vermeer family's fragile finances.

22 Huygens to Huygens (27 April 1679), in Van Leeuwenhoek (1948), p. 39.

23 Ford (1985); Ruestow (1983).

24 The reasons for this have been discussed by Alpers (1983), Ruestow (1995) and Cobb (2002b). For a discussion of the parallels between Vermeer and Leeuwenhoek, see Huerta (2003).

25 Leeuwenhoek to Grew (25 April 1679), in Van Leeuwenhoek (1948), p. 5.

26 Ford (1985).

27 Leeuwenhoek to Grew (18 March 1678), in Van Leeuwenhoek (1941), p. 329.

28 Dobell (1932), pp. 342–5.

29 Cited by Leeuwenhoek in a letter of December 1677; see Cole (1930), pp. 10–12. The original letter of Oldenburg's has been lost.

30 For a discussion of Leeuwenhoek's language, see Damsteegt (1982).

31 Swammerdam to Thévenot (28 April 1678), in Lindeboom (1975b), p. 108.

32 'I am not inclined to edit a book, although I am repeatedly advised to do so.' Leeuwenhoek to Grew (31 May 1678), in Van Leeuwenhoek (1941), p. 361.

33 Van Leewenheock (sic) (1677). A slightly different translation, together with attempts at identifying what Leeuwenhoek saw, is given in Dobell (1932), pp. 117–63.

34 The 'hot' taste of pepper and chilli is produced by chemicals known as capsaicins, which happen to stimulate the temperature receptors in the mouth. The resultant sensation of burning is, in fact, a gustatory illusion. Capsaicins are soluble in fat, not water, so drinking water or alcohol will not relieve the symptoms; the best thing to do is drink something fatty such as yoghurt or milk.

35 Ruestow (1984).

36 Beal (1676). Ironically, these bioluminescent phenomena were presumably caused by bacterial activity. Had someone turned a single-lens microscope on these glowing bits of meat, they would, like Leeuwenhoek, have seen tens of thousands of tiny creatures.

37 Birch (1756–7), vol. 3, p. 338.

38 Leewenhoecks (sic), (1677).

39 Birch (1756–7), vol. 3, p. 352.

40 Leeuwenhoek (1678). All quotations from this letter (including those in the previous paragraph) are taken from the translation that appears in Cole (1930), pp. 10–12. See also Leeuwenhoek to Brouncker (November 1677), in Van Leeuwenhoek (1941), pp. 279–99. There has been confusion as to the actual date of publication of Leeuwenhoek's letter, which appeared in the final issue of volume 12 of the *Philosophical Transactions* (number 142). The date given on the first page of this issue is 'For the Months of December, January, and February, 1678' (p. 1035), while volume 12 is generally described in libraries as being '1677–78'. At the time the new year was often still dated from 1 April, not 1 January, so this was the issue for December 1678–February 1679, as the printing information for number 142 makes clear (p. 1074). There are many accounts of the nature and impact of Leeuwenhoek's discovery; I have been particularly influenced by Ruestow (1983, 1995) and Roger (1963, 1997). See also Lindeboom (1982).

41 Birch (1756–7), vol. 3, pp. 493–4.

42 Leeuwenhoek to Grew (18 March 1678), in Van Leeuwenhoek (1941), p. 335.

43 Leeuwenhoek to Wren (22 January 1683), in Van Leeuwenhoek (1952), p. 11. It has been suggested that these 'vessels' were in fact tangles of fibrous matter (Lindeboom, 1982), although this does not really make things clearer.

44 Swammerdam to Thévenot (28 April 1678), in Lindeboom (1975b), p. 88.

45 Swammerdam (1758), pt 2, p. 147. Swammerdam warns the reader that his drawing is not entirely accurate: 'I have here represented it much shorter than it appears through the microscope' (Swammerdam, 1758, pt 2, p. 147). The study was announced to Thévenot in May 1678 (Lindeboom, 1975b, pp. 110–13), so must have been carried out in March or April. Swammerdam made no explicit connection between the 'small worms' he observed in dog semen and the 'particles' he saw in the cuttlefish. One reason for this may have been that although the cuttlefish sperm moved in water, they were far less urgent than the dog sperm. This is because in the cuttlefish, as in other molluscs, sperm are virtually inactive unless they are stimulated by chemicals secreted from an egg (Zatylny, Marvin, Gagnon and Henry, 2002).

46 Hartsoker (sic) (1678). For Bartholin, see Cole (1930), p. 15.

47 Huygens (1678).

48 Anonymous [N. Grew], 1678, p. 1043. Translation from Ruestow
    (1983), p. 193. According to the translations of this quotation in
    Roger (1963, p. 297; 1997, p. 238 – French and English, respec-
    tively) Grew states that semen 'imprints a vital contact onto the
    matter of the embryo, that is, the female egg' (Roger, 1997, p.
    238), which suggests that the egg contained the embryo prior to
    conception. The Latin original can also be interpreted in purely
    Aristotelian terms: 'quàm Vehiculum Spiritûs cujusdam summè volatilis
    ac animalis, & conceptioni, i.e. Ovo Faemineo contactum vitaliem impri-
    mentis.' (Anonymous [N. Grew], 1678, p. 1043.)

49 Leeuwenhoek to Hooke (12 November 1680), in Van
    Leeuwenhoek (1948), p. 329.

50 Leeuwenhoek to Grew (25 April 1679), in Van Leeuwenhoek
    (1948), pp. 3–35. Leeuwenhoek relates how he hired someone to
    castrate dogs for him so he could obtain the testicles. He eventu-
    ally learnt that this person had obtained one of the dogs from its
    owner on the pretext of de-fleaing it, and had taken the other dog
    off the street. Leeuwenhoek was not happy.

51 This translation is given in Ruestow (1983), p. 200. A duller trans-
    lation is given in Leeuwenhoek to Wren (22 January 1683), in Van
    Leeuwenhoek (1952), p. 11. The original, less precise, translation
    can be found in Leeuwenhoek (1683).

52 Leeuwenhoek to Wren (22 January 1683), in Van Leeuwenhoek
    (1952), p. 17.

53 Ibid. (22 January 1683), in Van Leeuwenhoek (1952) p. 13.

54 Leeuwenhoek to the Royal Society (30 March 1685), in Van
    Leeuwenhoek (1957), p. 201.

55 Leeuwenhoek to Grew (21 February 1679), in Van Leeuwenhoek
    (1941), p. 411.

56 Ibid. (18 March 1678), in Van Leeuwenhoek (1941), p. 359.

57 Leeuwenhoek to Wren (22 January 1683), in Van Leeuwenhoek
    (1952), p. 9.

58 Leeuwenhoek to Grew (18 March 1678), in Van Leeuwenhoek
    (1941), p. 343. See also: Leeuwenhoek to Hooke (12 November
    1680), in Van Leeuwenhoek (1948), p. 329; Leeuwenhoek to the
    Royal Society (30 March 1685), in Van Leeuwenhoek (1957), p.
    195.

59 Leeuwenhoek to the Royal Society (30 March 1685), in Van
    Leeuwenhoek (1957), p. 153.

60 Leeuwenhoek to Garden (19 March 1694), in Van Leeuwenhoek

(1976), pp. 58–9. The translation (p. 59) has 'figments' rather than 'imaginings', but both are acceptable.

61 Leeuwenhoek to the Royal Society (30 March 1685), in Van Leeuwenhoek (1957), p. 177. A different translation can be found in Leeuwenhoek (1685).

62 Leeuwenhoek to Wren (22 January 1683), in Van Leeuwenhoek (1952), p. 7.

63 Leeuwenhoek to the Royal Society (30 March 1685), in Van Leeuwenhoek (1957), p. 205.

64 Leeuwenhoek to Garden (19 March 1694), in Van Leeuwenhoek (1976), p. 57.

65 Garden to van Leeuwenhoek (24 August 1693), in Van Leeuwenhoek (1976), p. 39.

66 Ibid. (24 August 1693), in Van Leeuwenhoek (1976), pp. 43–5.

67 Ibid. (24 August 1693), in Van Leeuwenhoek (1976), p. 55.

68 Leeuwenhoek to Garden (19 March 1694), in Van Leeuwenhoek (1976), p. 57.

69 Leeuwenhoek to the Royal Society (30 March 1685), in Van Leeuwenhoek (1957), p. 169.

70 Leeuwenhoek to Garden (19 March 1694), in Van Leeuwenhoek (1976), p. 59.

71 B. H. [Benedictus Haan] to Leeuwenhoek (28 February 1694), in Van Leeuwenhoek (1976), p. 63.

72 Leeuwenhoek to Grew (25 April 1679), in Van Leeuwenhoek (1948), p. 27.

73 Ray (1693), pp. 60–61.

74 Quoted in Roger (1997), p. 253.

75 For an account of Leeuwenhoek's legacy, and how it disappeared, see Dobell (1932) and Ford (1985, 1991).

76 Harvey (1981), p. 183.

77 Hirai (2005).

78 Sherley (1672), p. 28.

CHAPTER 8: FROM GENERATION TO GENETICS
(pp. 220–53)

1 Bacon (1638), p. 149.

2 De Graaf (1672), p. 199. Translated in Jocelyn and Setchell (1972), p. 137.

3   Quoted in Leonhard (2002), p. 305.

4   Anonymous (1672b), p. 5000; Swammerdam (1672), p. 29.

5   Harvey (1981), p. 295.

6   Ibid. (1981), p. 295.

7   Leeuwenhoek gave a hint that he thought he might be able to explain sex determination, when he initially suggested that he could distinguish two kinds of 'spermatic animal' – male and female – before retracting this claim; see Leeuwenhoek to Wren (22 January 1683), in Van Leeuwenhoek (1952), p. 11. Sex determination is in fact remarkably varied in different organisms, involving different genetic and chromosomal effects. In mammals, birds and most insects, sex is determined by the possession of a single pair of chromosomes – the X and Y chromosomes in mammals and most insects, the W and Z chromosomes in birds and butterflies. In mammals males are XY and females XX, whereas in birds and butterflies males are ZZ and females WZ. Sex determination by specific chromosomes is not the rule, however. It can involve overall chromosome number (e.g. ants and bees), and even temperature effects (e.g. crocodiles). Some animals (and many plants) are hermaphrodite, while others can change their sex whether through genetic effects, the intervention of a parasite or in response to social or environmental changes. Sex determination by sex chromosomes in both mammals and plants is apparently a relatively recent evolutionary invention. Mittwoch (2000); D. Charlesworth (2004); B. Charlesworth (2004). Some of the ideas in the paragraphs that follow were suggested to me by Andrew Pyle, of Bristol University's Department of Philosophy.

8   Leeuwenhoek to Wren (16 July 1683), in Van Leeuwenhoek (1952), pp. 69–71.

9   Ibid. (16 July 1683), in Van Leeuwenhoek (1952), p. 69.

10 Seneca (1972), p. 289 [Book 3: 29, 3].

11 Fludd (1659), p. 78.

12 Preformation has been the subject of many scientific and historical analyses over the last century. See, for example, Punnett (1928); Cole (1930); Needham (1959); Adelmann (1965); Gasking (1967); Pinto-Correia (1997).

13 For a discussion of Hartsoeker's claims see Cole (1930).

14 Published in Hartsoeker (1694).

15 For a fascinating account of how this image was used and misused

over the next three hundred years, see Pinto-Correia (1997). See also Hill (1985).

16 See, for example, the unsigned review in *Le Journal des Sçavans*, 7 February 1695 (Anonymous, 1695).

17 Cole (1930), pp. 63–6.

18 Plantade (1670–1741), visited Paris and London, met Cassini and was influenced by Newton. On returning to Montpellier he became a respected physicist and astronomer, studying air pressure and various lunar and solar eclipses. In 1706 he was elected to the Société Royale des Sciences de Montpellier. He regularly climbed the Pic du Midi mountain to carry out his observations; aged eighty-one he collapsed and expired there. His dying words were 'Ah! All this is so beautiful!' Intriguingly, two years after publishing his notorious description of sperm, he wrote a fairy tale about little people, in which he appears to ridicule both spermist and ovist conceptions. See Tucker (2004).

19 Dalenpatius (1699). An English translation is given in Cole (1930), pp. 68–70.

20 Ibid. (1699), p. 552.

21 Ibid. (1699), p. 553; Cole (1930), p. 69.

22 Ibid. (1699), p. 554; Cole (1930), p. 69.

23 Leuvenhook (sic) (1699), p. 305.

24 Ibid. (1699), p. 306.

25 Drake (1707).

26 Drake's sister Judith may have written the section of this edition which dealt with generation – see Cohen (1997).

27 This and subsequent quotes from Drake (1707), vol. 1, p. 187.

28 Cole (1946), p. 120.

29 Cole (1946), p. 127. Latin original in Birch (1756–7), vol. 3, pp. 30–40. An English translation, with commentary, can be found in Cole (1946).

30 Reproduced in Adelmann (1965).

31 Adelmann (1965), p. 869. As Schrader (1674) pointed out, much earlier in the century, Joseph de Aromatari had made a similar suggestion, which went unnoticed at the time. In his book, which was dedicated to Swammerdam and Slade, Schrader published Wilhelm Langly's 1655 and 1667 observations on chick development (pp. 136–68 and pp. 168–81, respectively), and his own long commentary on the anatomical detail of Harvey's *De Generatione*. Some commentators have suggested that Langly put forward the

idea that women have eggs, but the only statement about the question to be found in Schrader's book is his reference in the Preface to Steno's 1667 declaration.

32 Adelmann (1965), p. 869.

33 Swammerdam (1685), p. 59.

34 Ibid. (1685), pp. 91–2.

35 Ibid. (1685), p. 80; Swammerdam (1758), pt 1, p. 92.

36 Swammerdam (1672), pp. 21–2. The Levi/Melchisedec analogy was taken directly from *Historia Insectorum Generalis* (Swammerdam, 1685, pp. 47–8), which was subsequently reproduced in *The Book of Nature* (Swammerdam, 1758, pt 1, p. 16). He did not use the Adam and Eve example in either of these two works. For a full discussion of Swammerdam's views about generation, see Ruestow (1985, 1995).

37 For discussions of the scientific and philosophical differences between the two concepts, see Roger (1997) and Bowler (1971).

38 Leewenhoeck (sic) (1685).

39 Ray (1691), p. 217.

40 Quoted in Fouke (1989), p. 368.

41 Jacob (1993), p. 57.

42 For a philosophical discussion of Malebranche's attitude to preformation and pre-existence, see Pyle (2003). The ideas of pre-existence and preformation have been subject to withering criticism by some historians of science, not only because they were wrong, but also because they are sometimes perceived to have contributed to the nearly two-century-long 'delay' in the development of a fully scientific understanding of the role of the component parts of generation. In the early 1930s, one of the earliest and most influential historians of this period of science, F. J. Cole, argued that preformation was 'a deadly obstruction' that had 'stopped the clock' of scientific progress through its 'long and evil reign' (Cole, 1930), while in the 1950s Joseph Needham tracked what he considered to be its 'cloven hoof' down the centuries (Needham, 1959). It is now clear not only that a series of factors contributed to the long domination of ovism and spermism, but also that as preformation formed the scientific consensus of the eighteenth and early nineteenth centuries, and generated further discoveries — one of the most important measures of the value of a scientific theory — it was in fact remarkably prolific.

43 Rose (1997); Morange (2001).

44 See, for example, Glass (1959).

45 This period in the history of biology has been intensively studied by historians. The most thorough account can be found in Roger (1997); versions of different aspects can be found in Cole (1930); Needham (1959); Gasking (1967); Farley (1977); Jacob (1993); Roe (1981, 2003); Pinto-Correia (1997). All of these works have been consulted for the extremely brief summary that follows.

46 Capanna (1999).

47 Roger (1981), p. 232.

48 Wristberg's comment can be found in Haller (1786), vol. 2, p. 209, n. 183.

49 The summary in the following paragraphs is based on Cole (1930); Gasking (1967); Short (1977); Farley (1982); Jacob (1993). Nuck tied off one of the 'horns' in the uterus of a bitch that had recently mated and in which, on dissecting, he had seen two 'eggs' in the ovaries. He sewed the dog back up, waited nearly three weeks, killed her and then showed that there were two embryos in the part of the uterus that had been tied off. Because he thought no semen got into the Fallopian tubes, Nuck argued that the embryos must have come from eggs that had been fertilised, rather than from the male's semen. See Short (1977).

50 Haighton (1797) and Cruikshank (1797). Haighton tied off one 'horn' of a rabbit's uterus at various times after copulation, and was able to show that mating induced ovulation in the rabbit (this was the point which De Graaf had mistakenly generalised to all mammals, including humans), while Cruikshank came close to identifying the mammalian egg when he repeated De Graaf's experiments.

51 The experiments were carried out in April/May 1827, and were published as *De Ovi Mammalium et Hominis Genesi* in January 1828 (but dated 1827). A facsimile of Von Baer's original Latin text, together with an introductory essay, can be found in Von Baer and Sarton (1931). A translation can be found in Von Baer (1956).

52 Quote taken from Von Baer's 1864 autobiography, translation given in Bernard Cohen's introduction to Von Baer (1956), p. 120.

53 Von Baer (1956), p. 149.

54 For contrasting views of this period of nineteenth-century science, see, for example, Harris (1999) and Sapp (2003).

55 This work was done by Oscar Hertwig, August Weismann and Hermann Fol.

56 See Stephen Jay Gould's classic study *Ontogeny and Phylogeny* (Gould, 1977). For an alternative interpretation, see Sapp (2003).
57 There is a huge literature on the origins of genetics. For this extremely brief summary I have consulted Bowler (1989); Jacob (1993); Gayon (1998); Sapp (2003); Carlson (2004).
58 Some internal parasites have incredibly complex life cycles which involve internal and external stages. These examples, which took a great deal of work to elucidate, initially provided ammunition for those who argued in favour of spontaneous generation.
59 For a discussion of the thorny question 'What is life?', current views of how life first appeared, and a fascinating explanation of why space exploration has recently focused on the search for extraterrestrial life, see Michel Morange's book, *La Vie expliquée?* (Morange, 2003), which will soon appear in English translation as *The Secret of Life*. For a discussion of the origin of life which emphasises the primordial importance of something resembling the cell, see Steven Rose's *Lifelines* (Rose, 1997).

BIBLIOGRAPHY

MANUSCRIPT ARCHIVES

Volumen Inscriptionum sive Catalogus Studiosurum Academiae Leÿdensis (volume 4, 1645–62), Leiden University Library.

Royal Society Archives, London. Minutes of Council, volume 1, 1668–81.

Royal Society Archives, London. Letter Book O.

Royal Society Archives, London. Journal Book Original, volume V.

PRINTED WORKS

'Espinasse, M. (1956) *Robert Hooke* (London: Heinemann).

Adelmann, H. B. (1942) *The Embryological Treatises of Hieronymus Fabricius of Aquapendente* (Ithaca, NY: Cornell University Press).

—— (1965) *Marcello Malpighi and the Evolution of Embryology* (Cornell: Cornell University Press).

—— (ed.) (1975) *The Correspondence of Marcello Malpighi* (London: Cornell University Press).

Alpers, S. (1983) *The Art of Describing: Dutch Art in the Seventeenth Century* (Chicago: University of Chicago Press).

Ankum, W. M., Houtzager, H. L., and Bleker, O. P. (1996) 'Reinier de Graaf (1641–1673) and the Fallopian tube', *Human Reproduction Update*, 2: 365–9.

Anonymous (1667a) 'An account of some books', *Philosophical Transactions*, 2: 534–5.

—— (1667b) 'An account of some books', *Philosophical Transactions*, 2: 628.

—— (1668) 'An account of some books', *Philosophical Transactions*, 3: 603–4.

—— (1669) 'Some observations concerning the organs of generation, made by Dr. Edmund King, a Fellow of the R. Society, and by Dr Regnerus de Graeff, physitian in Holland; which later occasioned the publishing of the former', *Philosophical Transactions*, 4: 1043–7.

—— (1670) 'An Accompt of some books', *Philosophical Transactions*, 6: 2078–81.

—— (1671a) 'An Intimation of divers Philosophical particulars, now undertaken and consider'd by several Ingenious and Learned men; here inserted to excite others to joyn with them in the same or like Attempts and Observations', *Philosophical Transactions*, 6 (74): 2216–19.

—— (1671b) 'An Accompt of some Books', *Philosophical Transactions*, 6: 2136–7.

—— (1671c) 'An Accompt of some Books', *Philosophical Transactions*, 6: 2210–14.

—— (1672a) 'An Account of what hath been of late observed by Dr. Kerkringius concerning Eggs to be found in all sorts of Females', *Philosophical Transactions*, 7: 4018–26.

—— (1672b) 'An accompt of some books', *Philosophical Transactions*, 7: 4095–5002.

—— (1679a) *Catalogus Musei Instructissimi . . . Johannes Jacobus Swammerdammius Pharmacop. dum viveret vigilantissimus* (n.p.: d'Erfgenamen).

—— (1679b) 'Gaspari Bartholini Th. Filii De Ovariis mulierum & generationis historia Epistola Anatomica', *Journal des Sçavans*, 6 March 1679, pp. 63–4.

—— (1694) *Bibliotheca Thevenotiana* (Paris: Delaulne).

—— (1695) 'Essay de Dioptrique. Par N. Hartoseker', *Journal des Sçavans*, 7 February 1695, pp. 65–9.

—— (1878) 'Notice sur Regnier de Graaf', in Reinier de Graaf, *L'Instrument de Molière* (Paris: Morgand et Fatout).

—— [Jean Gallois] (1672) 'Discours des oeufs dont il est parlé dans le livre de M. Kerckring', *Journal des Sçavans*, 21 March 1672, pp. 65–8.

—— [Henry Oldenburg?] (1672) 'M. Joannis Sammerdami MD Uteri

Muliebris Fabrica', *Philosophical Transactions*, 7: 4098–5001.

—— [N. Grew] (1678) 'Auctoris ad observatorem responsum', *Philosophical Transactions*, 12: 1043.

Aubrey, J. (2000) *Brief Lives* (Harmondsworth: Penguin).

Bacon, F. (1638) *Historie Naturall and Experimentall, of Life and Death. Or of the Prolongation of Life* (London: Lee & Mosley).

Baldwin, M. (1995) 'The snakestone experiments: An early modern medical debate', *Isis*, 86: 394–418.

Barles, L. (1674) *Les Nouvelles découvertes sur les organes des femmes, servans à la generation* (Lyon: Esprit Vitalis).

Bartholini, T. (1680) 'De ovariis mulierum', in T. Bartholini (ed.), *Acta Medica & Philosophica Hafniensa* (Hafniae, Hauboldi), vol. 5, pp. 33–54.

Baumann, E. D. (1949) *François dele Boe Sylvius* (Leiden: Brill).

Beal, J. (1676) 'Two instances of something remarkable in shining flesh, from Dr. J. Beal of Yeavel in Somersetshire, in a letter to the publisher', *Philosophical Transactions*, 11: 599–603.

Bennet, J., Cooper, M., Hunter, M., and Jardine, L. (2003) *London's Leonardo: The Life and Works of Robert Hooke* (Oxford: Oxford University Press).

Bergin, J. (2001) *The Seventeenth Century* (Oxford: Oxford University Press).

Bernardi, W., and Guerrini, L. (eds) (1999) *Francesco Redi: Un Protagonista della Scienza Moderna. Documenti, Esperimenti, Immagini* (Florence: Olschki).

Biagioli, M. (1992) 'Scientific revolution, social bricolage, and etiquette', in R. Porter and M. Teich (eds) *The Scientific Revolution in National Context* (Cambridge: Cambridge University Press), pp. 11–54.

—— (1995) 'Le prince et les savants: La civilité scientifique au 17e siècle', *Annales – Histoire Sciences Sociales*, 50: 1417–55.

Bignami, G. F. (2000) 'The microscope's coat of arms', *Nature*, 405: 999.

Birch, T. (1756–7) *The History of the Royal Society of London for Improving of Natural Knowledge* (London: Millar), 4 volumes [1968 Facsimile edition (Hildesheim: Olms)].

Blackledge, C. (2003) *The Story of V* (London: Weidenfeld & Nicolson).

Boerhaave, H. (1719) *A Method of Studying Physick* (London: Rivington, Creake and Sackfield).

—— (1744) *Praelectiones Academicae* (Amsterdam: n.p.).

Bonnet, C. (1783) *Oeuvres d'histoire naturelle et de philosophie*. Tome quinzième: *La Palingénésie philosophique*, Parts 1–XI (Neuchatel: Fauche).

Borel, P. (1656) *Observationum Microscopicarum Centuria* (The Hague: Vlacq).

Boschiero, L. (2002) 'Natural philosophizing inside the late seventeenth-century Tuscan court', *British Journal for the History of Science*, 35: 383–410.

—— (2003) 'Natural philosophical contention inside the Accademia del Cimento: the properties and effects of heat and cold', *Annals of Science*, 60: 329–49.

Bourchenin, P.-D. (1884) *De Tanaquilli Fabri: Vita et Scriptis* (Paris: Grasset).

—— (1887) *Etude sur les Académies Protestantes* (Paris: Grasset).

Boutibonnes, P. (1994) *Antoni van Leeuwenhoek 1632–1723: L'Exercice du regard* (Paris: Belin).

Bowler, P. J. (1971) 'Preformation and pre-existence in the seventeenth century: A brief analysis', *Journal of the History of Biology*, 4: 221–44.

—— (1989) *The Mendelian Revolution* (London: Athlone Press).

Boyle, R. (2000) 'Fragments of Boyle's Essay on Spontaneous Generation', in M. Hunter and E. B. Davis (eds) *The Works of Robert Boyle*, vol. 13: *Unpublished Writings, 1645–c. 1670* (London: Pickering & Chatto), pp. 270–90.

Bray, F. (1984) 'Part II: Agriculture', in J. Needham (ed.) *Science and Civilisation in China*: Volume 6: *Biology and Biological Technology* (Cambridge: Cambridge University Press).

Brockbank, W., and Corbett, O. R. (1954) 'De Graaf's "Tractatus de Clysteribus"', *Journal of the History of Medicine and Allied Sciences*, 9: 174–90.

Brown, H. (1934) *Scientific Organizations in Seventeenth-Century France* (Baltimore: Williams & Wilkins).

Cadden, J. (1993) *Meanings of Sex Difference in the Middle Ages: Medicine, Science and Culture* (Cambridge: Cambridge University Press).

Capanna, E. (1999) 'Lazzaro Spallanzani: At the roots of modern biology', *Journal of Experimental Zoology*, 285: 178–96.

Carlson, E. O. (2004) *Mendel's Legacy: The Origin of Classical Genetics* (Cold Spring Harbor, NY: Cold Spring Harbor Laboratory Press).

Catchpole, H. R. (1940) 'Regnier de Graaf 1641–1673', *Bulletin of the History of Medicine and Allied Sciences*, 8: 1261–1300.

Cetto, A. (1958) 'An unknown portrait of Regnier de Graaf', *CIBA Symposium*, 5: 208–11.

Chapman, A. (2005) *England's Leonardo: Robert Hooke and the Seventeenth-Century Scientific Revolution* (Bristol: Institute of Physics).

Charlesworth, B. (2004) 'Evolutionary biology: On chromosomes and sex determination', *Current Biology*, 14: R745.

Charlesworth, D. (2004) 'Plant evolution: Modern sex chromosomes', *Current Biology*, 14: R271–R273.

Cherry, R. (1989) 'Sericulture', *Bulletin of the Entomological Society of America*, 35: 83–4.

Cioni, R. (1962) *Niels Stensen Scientist-Bishop* (New York: Kenedy & Sons).

Clarck (sic) T. (1668) 'A Letter, Written to the Publisher by the Learned and Experienced Dr Timothy Clarck One of His Majesties Physitians in Ordinary, Concerning Some Anatomical Inventions and Observations, Particularly the Origin of the Injection into Veins, the Transfusion of Bloud, and the Parts of Generation', *Philosophical Transactions*, 3: 172–82.

Cobb, M. (2000) 'Reading and writing The Book of Nature: Jan Swammerdam (1637–1680)', *Endeavour*, 24: 122–8.

—— (2002a) 'Exorcizing the animal spirits: Jan Swammerdam and nerve function', *Nature Reviews: Neuroscience*, 5: 395–400.

—— (2002b) 'Malpighi, Swammerdam and the colourful silkworm: Replication and visual representation in early modern science', *Annals of Science*, 59: 111–47.

Cochrane, E. (1973) *Florence in the Forgotten Centuries 1527–1800* (London: University of Chicago Press).

Cohen, E. (1997) '"What the women at all times would laugh at": Redefining equality and difference, circa 1660–1760', *Osiris*, 12: 121–42.

Cohen, H. F. (1994) *The Scientific Revolution* (London: University of Chicago Press).

Cole, F. J. (1921) 'The history of anatomical injections', in C. Singer (ed.) *Studies in the History and Method of Science* (Oxford: Clarendon Press), vol. 2, pp. 285–343.

—— (1930) *Early Theories of Sexual Generation* (Oxford: Clarendon Press).

—— (1944) *A History of Comparative Anatomy* (London: Macmillan).

—— (1946) 'Dr. William Croune on Generation', in M. F. Ashley Montagu (ed.) *Studies and Essays in the History of Science and Learning* (New York: Schuman), pp. 113–35.

Cook, H. J. (1994) *Trials of an Ordinary Doctor* (London: Johns Hopkins University Press).

—— (2002) 'Time's bodies: Crafting the preparation and preservation of naturalia', in P. H. Smith and P. Findlen (eds) *Merchants & Marvels: Commerce, Science and Art in Early Modern Europe* (London: Routledge), pp. 223–47.

Cowles, T. (1934) 'Dr Henry Power's poem on the Microscope', *Isis*, 21: 71–80.

Croxatto, H.B. (2002) 'Physiology of gamete and embryo transport through the Fallopian tube', *Reproductive Biomedicine Online*, 4: 160–9.

Cruikshank, W. (1797) 'Experiments in which, on the third day after impregnation, the ova of rabbits were found in the fallopian tubes; and on the fourth day after impregnation in the uterus itself; with the first appearances of the foetus', *Philosophical Transactions of the Royal Society of London*, 87: 197–214.

Cutler, A. (2003) *The Seashell on the Mountaintop* (London: Heinemann).

Dalenpatius (1699) 'Extrait d'une lettre de M. Dalenpatius à l'auteur de ces Nouvelles, concernant une découverte curieuse, faite par le moyen du microscope', *Nouvelles de la République des Lettres*, May 1699, pp. 552–4.

Damsteegt, B. C. (1982) 'Language and Leeuwenhoek', in L. C. Palm and H. A. M. Snelders (eds) *Antoni van Leeuwenhoek 1632–1723* (Amsterdam: Rodopi), pp. 13–28.

Daston, L., and Park, K. (1998) *Wonders and the Order of Nature, 1150–1750* (New York: Zone Books).

Dawmarn, F. (2003) 'New light on Dr Thomas Moffet: The triple roles of an early modern physician, client and patronage broker', *Medical History*, 47: 3–22.

De Baar, M. (1994) 'Transgressing gender codes: Anna Maria van Schurman and Antoinette Bourignon as contrasting examples', in E. Kloek, N. Teeuwen and M. Huisman (eds) *Women of the Golden Age* (Hilversum: Verloren), pp. 143–52.

De Crousaz-Crétet, P. (1922) *Paris sous Louis XIV: La Vie privée et la vie professionnelle* (Paris: Plon).

De Diemerbroeck, I. (1694) *The Anatomy of Human Bodies, Comprehending the Most Modern Discoveries and Curiosities in That Art* (London: Whitwood).

De Graaf, R. (1664) *Disputatio Medica de Natura et Usu Succi Pancreatici* (Leiden: Hack).

—— (1668) *De Vivorum Organis Generationi inservientibus de Clysteribus et de usu Siphonis in Anatomia* (Leiden: Hack).

—— (1671) 'Epistola ad virum clarissimum D. Lucam Schacht', in R.

de Graaf *Tractatus Anatomico-Medicus de Succi Pancreatici Natura & Usu* (Leiden: Hack), pp. 209–16.

—— (1672) *De Mulierum Organis Generationi Inservientibus* (Leiden: Hack).

—— (1673) *Partium Genitalium Defensio* (Leiden: Hack).

—— (1699) *Histoire anatomique des parties genitales de l'homme et de la femme, qui servent à la generation, traduit par monsieur NPDM* (Basle: Kong).

—— (1878) *L'instrument de Molière: traduction de traité De Clysteribus de Regnier de Graaf (1668)* (Paris: Morgand et Fatout).

De Monconys, B. (1666) *Journal des voyages de monsieur de Monconys. II* (Lyon: Boissat & Remeus).

De Thévenot, J. (1665) *Relation d'un voyage au levant* (Paris: Iolly).

De Wiquefort, A. (1673) *Advis fidelle aux veritables Hollandois. Touchant ce qui s'est passé dans les villages de Bodegrave & Swammerdam, & les cruautés inoüies, que les François y ont exercées* (The Hague: Steucker).

Dekker, R. M. (1999) 'Sexuality, elites and court life in the late seventeenth century: The diaries of Constantijnn Huygens, Jr.', *Eighteenth-Century Life*, 23: 94–109.

—— (2001) *Humour in Dutch Culture of the Golden Age* (Palgrave: Basingstoke).

Dennis, M. A. (1989) 'Graphic understanding: Instruments and interpretation in Robert Hooke's *Micrographia*', *Science in Context*, 3: 309–64.

Dibon, P. (1963) *Le Voyage en France des étudiants Néerlandais au XVIIème siècle* (The Hague: Martinus Nijhoff).

Digby, K. (1644) *Two Treatises In the One of Which the Nature of Bodies, in The Other, the Nature of Mans Soule is Looked Into in Way of Discovery of the Immortality of Reasonable Soules* (Paris: Blaizot).

Dobell, C. (1932) *Antony van Leeuwenhoek and his 'Little Animals'* (London: Constable) [1960 Facsimile edition (New York: Dover)].

Drake, J. (1707) *Anthropologia Nova; or, A New System of Anatomy* (London: n.p.).

Duchêne, R. (2004) *Etre femme au temps de Louis XIV* (Paris: Perrin).

Duchess of Newcastle [Margaret Cavendish] (1666) *Observations upon Experimental Philosophy* (London: Maxwell).

—— (1671) *Natures Pictures Drawn by Fancies Pencil To the Life* (London: Maxwell).

Engel, H. (1986) *Hendrik Engel's Alphabetical List of Dutch Zoological Cabinets and Menageries* (Amsterdam: Rodopi).

Farley, J. (1977) *The Spontaneous Generation Controversy from Descartes to Oparin* (Baltimore: Johns Hopkins University Press).

—— (1982) *Gametes and Spores: Ideas about Sexual Reproduction 1750–1914* (London: Johns Hopkins University Press).

Findlen, P. (1993) 'Controlling the experiment: rhetoric, court patronage and the experimental method of Francesco Redi', *History of Science*, 31: 35–64.

—— (1994) *Possessing Nature: Museums, Collecting and Scientific Culture in Early Modern Italy* (London: University of California Press).

—— (ed.) (2004) *Athanasius Kircher: The Last Man Who Knew Everything* (London: Routledge).

Fischer-Homberger, E. (2001) *Harvey's Troubles with the Egg* (Sheffield: European Association for the History of Medicine and Health).

Fissell, M. E. (2004) *Vernacular Bodies: The Politics of Reproduction in Early Modern England* (Oxford: Oxford University Press).

Fludd, R. (1659) *Mosaicall Philosophy Grounded Upon the Essentiall Truth, or Eternal Sapience* (London: Moseley).

Fontana, F. (1646) *Novae Coelestium, Terrestriumque Rerum Observationes* (Naples: Gaffarum).

Foote, E. T. (1969) 'Harvey: Spontaneous generation and the egg', *Annals of Science*, 25: 139–63.

Ford, B. J. (1985) *Single Lens: The Story of the Simple Microscope* (New York: Harper & Row).

—— (1991) *The Leeuwenhoek Legacy* (Bristol: Biopress).

—— (2000) 'Robert Hooke and the Royal Society, by Richard Nichols', *Endeavour*, 24: 45–6.

Fouke, D. C. (1989) 'Mechanical and "organical" models of seventeenth-century explanations of biological reproduction', *Science in Context*, 3: 365–81.

Fournier, M. (1996) *The Fabric of Life: Microscopy in the Seventeenth Century* (London: Johns Hopkins University Press).

Freedberg, D. (1991) 'Science, commerce and art: neglected topics at the junction of history and art history', in D. Freedberg and J. de Vries (eds) *Art in History, History in Art* (Santa Monica: Getty), pp. 377–428.

—— (2002) *The Eye of the Lynx* (London: Chicago University Press).

French, R. (2003) *Medicine Before Science* (Cambridge: Cambridge University Press).

Furth, C. (1987) 'Concepts of pregnancy, childbirth and infancy in Ch'ing dynasty China', *Journal of Asian Studies*, 46: 7–35.

—— (1999) *A Flourishing Yin: Gender in China's Medical History, 960–1665* (London: University of California Press).

Gadbury, J. (1691) 'A diary of the weather for XXI years together, exactly observed in London, with sundry observations thereon', in J. Gadbury, *Nauticum Astrologicum: or The Astrological Seaman* (London: Street), pp. 129–245.

Galen (1968) *Galen on the Usefulness of the Parts of the Body* (Cornell: Cornell University Press).

Garmann, C. (1672) *Homo ex Ovo, sive De Ovo Humano Dissertatio* (Chemnitz: Sumptibus Authoris).

Gasking, E. B. (1967) *Investigations into Generation 1651–1828* (London: Hutchinson).

Gayon, J. (1998) *Darwinism's Struggle for Survival: Heredity and the Hypothesis of Natural Selection* (Cambridge: Cambridge University Press).

Ginzburg, C. (1980) *The Cheese and the Worms: The Cosmos of a Sixteenth-Century Miller* (Baltimore: Johns Hopkins University Press).

Glass, B. (1959) 'The germination of the idea of biological species', in B. Glass, O. Temkin and W. L. Strauss Jr. (eds) *Forerunners of Darwin: 1745–1859* (Baltimore: Johns Hopkins University Press), pp. 30–48.

Goedaert, J. (1662–8) *Metamorphosis et Historia Naturalis Insectorum* (3 vols) (Middelburg: Jacobum Fiernsium).

—— (1682) *Of Insects Johannes Goedaert; Done into English and Methodized with the Addition of Notes* (London: M. L.).

Gonzalès, J. (1996) *Histoire naturelle et artificielle de la procréation* (Paris: Bordas).

Gottdenker, P. (1979) 'Francesco Redi and the fly experiments', *Bulletin of the History of Medicine*, 53: 575–92.

Gould, S. J. (1977) *Ontogeny and Phylogeny* (London: Belknap).

—— (1987) *Time's Arrow, Time's Cycle* (London: Cambridge University Press).

—— (2004) 'Father Athanasius on the Isthmus of a Middle State: Understanding Kircher's Paleontology', in P. Findlen (ed.) *Athanasius Kircher: The Last Man Who Knew Everything* (London: Routledge), pp. 207–37.

Gowing, L. (2003) *Common Bodies: Women, Touch and Power in Seventeenth-Century England* (London: Yale University Press).

Graham, P. W. (1978) 'Harvey's *De motu cordis*: The rhetoric of science and the science of rhetoric', *Journal of the History of Medicine and Allied Sciences*, 33: 469–76.

Greaves, R. L. (1969) 'Puritanism and science: The anatomy of a controversy', *Journal of the History of Ideas*, 30: 345–68.

Grew, N. (1686) *Musaeum Regalis Societatis: or, a Catalogue and Description of the Natural and Artificial Rarities Belonging to the Royal Society, and Preserved at Gresham Colledge* (London: Mathu).

Gribbin, J. (2005) *The Fellowship: The Story of a Revolution* (London: Allen Lane).

Grmek, M. D. (1990) *La Première Révolution biologique* (Paris, Payot).

Guyénot, E. (1941) *L'Evolution de la pensée scientifique: Les Sciences de la vie au XVIIe et XVIIIe siècles* (Paris: Albin Michel).

Haighton, J. (1797) 'An experimental inquiry concerning animal impregnation', *Philosophical Transactions of the Royal Society of London*, 87: 159–96.

Hale, J. R. (2001) *Florence and the Medici* (London: Phoenix).

Hall, A. R., and Hall, M. (eds) (1965) *The Correspondence of Henry Oldenburg*, vol. 1 (Madison: University of Wisconsin Press).

—— (eds) (1966a) *The Correspondence of Henry Oldenburg*, vol. 2 (Madison: University of Wisconsin Press).

—— (eds) (1966b) *The Correspondence of Henry Oldenburg*, vol. 3 (Madison: University of Wisconsin Press).

—— (eds) (1967) *The Correspondence of Henry Oldenburg*, vol. 4 (Madison: University of Wisconsin Press).

—— (eds) (1968) *The Correspondence of Henry Oldenburg*, vol. 5 (Madison: University of Wisconsin Press).

—— (eds) (1969) *The Correspondence of Henry Oldenburg*, vol. 6 (Madison: University of Wisconsin Press).

—— (eds) (1970) *The Correspondence of Henry Oldenburg*, vol. 7 (Madison: University of Wisconsin Press).

—— (eds) (1971) *The Correspondence of Henry Oldenburg*, vol. 8 (Madison: University of Wisconsin Press).

—— (eds) (1973) *The Correspondence of Henry Oldenburg*, vol. 9 (Madison: University of Wisconsin Press).

Hall, A. R. (1989) 'The Leeuwenhoek Lecture, 1988: Antoni van Leeuwenhoek 1632–1723', *Notes and Records of the Royal Society of London*, 43: 249–73.

Hall, M. B. (1966) 'Introduction', in Henry Power, *Experimental Philosophy* (New York: Johnson Reprint Company).

—— (1991) *Promoting Experimental Learning: Experiment and the Royal Society 1660–1727* (Cambridge: Cambridge University Press).

—— (2002) *Henry Oldenburg: Shaping the Royal Society* (Oxford: Oxford University Press).

Harris, H. (1999) *The Birth of the Cell* (London: Yale University Press).

—— (2001) *Things Come To Life* (Oxford: Oxford University Press).

Harth, E. (1983) *Ideology and Culture in Seventeenth-Century France* (London: Cornell University Press).

Hartsoeker, N. (1694) *Essai de dioptrique* (Paris: Anisson).

Hartsoker (sic), N. (1678) 'Extrait d'une lettre de M. Nicolas Hartsoker écrite à l'auteur du journal touchant la maniere de faire les nouveaux microscopes, don't il a esté parlé dans le journal il y a quelques jours', *Journal des Sçavans*, 22 August 1678, pp. 355–6.

Harvey, W. (1651) *Exercitationes de Generatione Animalium* (Amsterdam: Jansson).

—— (1653) *Anatomical Exercitations, Concerning the Generation of Living Creatures* (London: Young).

—— (1981) *Disputations Touching the Generation of Animals* (translated by Gweneth Whitteridge) (Oxford: Blackwell).

Harwood, J. T. (1989) 'Rhetoric and graphics in *Micrographia*', in M. Hunter and S. Schaffer (eds) *Robert Hooke: New Studies* (Woodbridge: Boydell Press), pp. 119–48.

Hett, F. P. (ed.) (1932) *The Memoirs of Sir Robert Sibbald (1641–1722)* (London: Oxford University Press).

Hibbert, C. (1974) *The Rise and Fall of the House of Medici* (London: Allen Lane).

Highmore, N. (1651) *The History of Generation* (London: Martin).

Hill, C. (1964) 'William Harvey and the idea of monarchy', *Past & Present*, 27: 54–72.

—— (1965) 'William Harvey (no parliamentarian, no heretic) and the idea of monarchy', *Past & Present*, 31: 97–103.

Hill, K. A. (1985) 'Hartsoeker's homunculus: A corrective note', *Journal of the History of the Behavioral Sciences*, 21: 178–9.

Hirai, H. (2005) *Le Concept de semence dans les théories de la matière à la Renaissance* (Turnhout: Brepols).

Hirschfield, J. M. (1981) *The Académie Royale des Sciences* (New York: Arno Press).

Hodierna, G. (1644) *L'Occio della Mosca* (Palermo: n.p.).

Hoogewerff, G. J. (1919) *De Twee Reizen van Cosimo de' Medici Prins van Toscane door de Nederlanden (1667–1669)* (Amsterdam: Müller).

Hook (sic) (1667) 'An account of an experiment made by Mr. Hook, of preserving animals alive by blowing through their lungs with bellows', *Philosophical Transactions*, 2: 539–40.

Hooke, R. (1665) *Micrographia or Some Physiological Descriptions of Minute*

Bodies Made by Magnifying Glasses with Observations and Inquiries Thereupon (London: Martyn and Allestry).

Houliston, V. H. (1989a) 'Sleepers awake: Thomas Moffet's challenge to the College of Physicians of London, 1584', *Medical History*, 33: 235–46.

—— (ed.) (1989b) *The Silkeworms and their Flies – Thomas Moffet* (Binghamton, NY: Renaissance English Text Society).

Howell, J. (1650) *Epistolae Ho-Elianae* (London: Moseley).

Hsia, R. P. and Van Nierop, H. F. K. (2002) *Calvinism and Religious Toleration in the Dutch Golden Age* (Cambridge: Cambridge University Press).

Huerta, R. D. (2003) *Giants of Delft* (Lewisburg: Bucknell University Press).

Hufton, O. (1995) *The Prospect Before Her: A History of Women in Western Europe. Vol. 1: 1500–1800* (London: HarperCollins).

Huygens, C. (1678) 'Extrait d'une lettre de M. Huygens de l'Acadèmiie R. des Sciences à l'auteur du journal, touchant une nouvelle manière de microscope qu'il a apporté de Hollande', *Journal des Sçavans*, 8 August 1678, pp. 331–2.

—— (1893) *Oeuvres Complètes de Christiaan Huygens*, vol. 5 (The Hague: Nijhoff).

—— (1974a) *Oeuvres Complètes de Christiaan Huygens*, vol. 3 (Amsterdam: Swets & Zeitlinger).

—— (1974b) *Oeuvres Complètes de Christiaan Huygens*, vol. 4 (Amsterdam: Swets & Zeitlinger).

—— (1977) *Oeuvres Complètes de Christiaan Huygens*, vol. 5 (Amsterdam: Swets & Zeitlinger).

I. W. [John Wilkins] (1648) *Mathematicall Magick or, The Wonders that May be Performed with Mechanicall Geometry* (London: Gellibrand).

Inwood, S. (2002) *The Man Who Knew Too Much* (London: Macmillan).

Israel, J. (1995) *Dutch Primacy in World Trade* (Oxford: Oxford University Press).

—— (1998) *The Dutch Republic. Its Rise, Greatness and Fall 1477–1806* (Oxford: Clarendon Press).

Jaco, J. (ed.) (1995) *Paracelsus: Selected Writings* (Princeton, NJ: Princeton University Press).

Jacob, F. (1993) *The Logic of Life: A History of Heredity* (Princeton, NJ: Princeton University Press).

Jardine, L. (1999) *Ingenious Pursuits: Building the Scientific Revolution* (London: Little, Brown).

—— (2002) *On a Grander Scale: The Outstanding Career of Sir Christopher Wren* (London: HarperCollins).

—— (2003) *The Curious Life of Robert Hooke* (London: HarperCollins).

Jocelyn, H. D., and Setchell, B. P. (1972) 'Regnier de Graaf on the human reproductive organs', *Journal of Reproduction and Fertility*, Supplement 17: 5.

Jones, J. R. (1996) *The Anglo-Dutch Wars of the Seventeenth Century* (London: Longman).

Kardel, T. (1994a) 'Steno: Life, science, philosophy', *Acta Historica Scientiarum Naturalium et Medicinalium*, 42: 1–159.

—— (1994b) 'Stensen's Myology in historical perspective', *Transactions of the American Philosophical Society*, 84: 1–57.

Keller, E. (1999) 'Making up for losses: The workings of gender in William Harvey's *de Generatione Animalium*', in S. C. Greenfield and C. Barash (eds) *Inventing Maternity: Politics, Science, and Literature 1650–1865* (Lexington: University Press of Kentucky), pp. 34–56.

Kerckring, T. (1670) *Spiculegium Anatomicum* (Amsterdam: Frisii).

—— (1671) *Anthropogeniae Ichnographia* (Amsterdam: n.p.).

Keynes, G. (1966) *The Life of William Harvey* (Oxford: Clarendon Press).

Kidd, M. and Modlin, I. M. (1999) 'The Luminati of Leiden: From Bontins to Boerhaave', *World Journal of Surgery*, 23: 1307–14.

Klein, M. (1970–80) 'Regnier de Graaf', in C. Gillespie (ed.) *Dictionary of Scientific Biography* (New York: Scribner).

Knellwolf, C. (2001) 'Robert Hooke's *Micrographia* and the aesthetics of empiricism', *The Seventeenth Century*, 16: 177–200.

Knoefel, P. K. (1988) *Francesco Redi on Vipers* (Leiden: Brill).

Laqueur, T. (1990) *Making Sex: Body and Gender from the Greeks to Freud* (Harvard: Harvard University Press).

Latham, R. (ed.) (1985) *The Shorter Pepys* (London: Unwin Hyman).

Leeuwenhoek, A. (1678) 'Observationes D. Anthonii Lewenhoeck, de natis è semini genitali animalcules', *Philosophical Transactions*, 12: 1040–3.

—— (1683) 'An abstract of a Letter from Mr. Anthony Leewenhoeck writ to Sir C.W. Jan. 22. 1682/3 from Delft', *Philosophical Transactions*, 13: 74–81.

—— (1685) 'An abstract of a letter of Mr. Leeuwenhoeck Fellow of the R. Society dated March 30th 1685 to the R.S. concerning generation by an insect', *Philosophical Transactions*, 15: 1120–34.

Leewenhoeck (sic) (1685) 'An Extract of a Letter from Mr Anthony Leewenhoeck F. of the R.S. to a S. of the R. Society, Dated from

Delf, January 5th. 1685, Concerning the Salts of Wine and Vinegar', *Philosophical Transactions*, 15: 963–79.

Leewenhoecks (sic) (1677) 'Monsieur Leewenheocks letter to the publisher, wherein some account is given of the manner of his observing so great a number of little animals in divers sorts of water, as was deliver'd in the next foregoing Tract: English'd out of Dutch', *Philosophical Transactions*, 12: 844–6.

Leonhard, K. (2002) 'Vermeer's pregnant women. On human generation and pictorial representation', *Art History*, 25: 293–318.

Leoniceni, J. (1673) *Metamorphosis Aesculapii et Apollinis Pancreatici* (Leiden: Bonnona).

Leuvenhook (sic) (1699) 'Part of a letter from Mr Leuvenhook, dated June 9th 1699, concerning the animalcula in semine humano &c.', *Philosophical Transactions*, 21: 301–8.

Libavius, A. (1661) 'Observatio Bombicum, historia singularis', in Joh. Jonstoni (ed.) *Thaumatographia Naturalis* (Amsterdam: Jansson), pp. 379–406.

Lindeboom, G. A. (1973) *Reinier de Graaf: Leven en Werken* (Delft: Elmar).

—— (1975a) 'A short biography of Jan Swammerdam (1637–1680)', in G. A. Lindeboom (ed.) *The Letters of Jan Swammerdam to Melchisedec Thévenot* (Amsterdam: Swets & Zeitlinger), pp. 1–27.

—— (ed.) (1975b) *The Letters of Jan Swammerdam to Melchisedec Thévenot* (Amsterdam: Swets & Zeitlinger).

—— (1982) 'Leeuwenhoek and the problem of sexual reproduction', in L. C. Palm and H. A. M. Snelders (eds) *Antoni van Leeuwenhoek 1632–1723* (Amsterdam: Rodopi), pp. 129–52.

Lister, M. (1671) 'A considerable accompt touching *vegetable excrescencies*', *Philosophical Transactions*, 6: 2254–7.

—— (1699) *A Journey to Paris in the Year 1698* (London: Tonson) [1967 Fascimile edition (London: University of Illinois Press)].

Litchfield, R. B. (1986) *Emergence of a Bureaucracy: The Florentine Patricians 1530–1790* (Guildford: Princeton University Press).

Lunsingh Scheurleer, Th. H., and Posthumus Meyhes, G. H. M. (eds) (1975) *Leiden University in the Seventeenth Century. An Exchange of Learning* (Leiden: Brill).

Lux, D. S. (1989) *Patronage and Royal Science in Seventeenth-Century France: The Académie de Physique in Caen* (London: Cornell University Press).

—— and Cook, H. S. (1998) 'Closed circles or open networks?:

Communicating at a distance during the scientific revolution', *History of Science*, 36: 179–211.

Luyendijk-Elshout, A. M. (1965) 'Introduction', in F. Ruysch, *Dilucidatio Valvorum in Vasis Lymphaticis et Lacteis* (Facsimile) (Nieuwkoop: De Graaf), pp. 7–49.

—— (1970) 'Death enlightened: A study of Frederik Ruysch', *Journal of the American Medical Association*, 212: 121–6.

Malpighi, M. (1669) *Dissertatio Epistolica de Bombyce* (London: Martin & Allestry).

Mauries, P. (2002) *Cabinets of Curiosities* (London: Thames & Hudson).

Mayerne, T. (1658) 'The Epistle Dedicatory' to T. Muffet, *The Theater of Insects*, in Edward Topsel, *The History of Four-footed Beasts and Serpents* (London: Sawbridge, Williams and Johnson).

Mayhew, R. (2004) *The Female in Aristotle's Biology* (London: University of Chicago Press).

McCartney, E. S. (1920) 'Spontaneous generation and kindred notions in Antiquity', *Transactions and Proceedings of the American Philological Association*, 51: 101–15.

McClaughlin, T. (1974) 'Une lettre de Melchisédech Thévenot', *Revue d'Histoire des Sciences*, 27: 123–6.

—— (1975) 'Sur les rapports entre la Compagnie de Thévenot et l'Académie royale des Sciences', *Revue d'Histoire des Sciences*, 28: 235–42.

McKeon, R. M. (1965) 'Une lettre de Melchisédech Thévenot sur les débuts de l'Académie royale des Sciences', *Revue d'Histoire des Sciences*, 18: 2–3.

Mendelsohn, J. A. (1992) 'Alchemy and politics in England 1649–1665', *Past & Present*, 135: 30–78.

Meyer, A. W. (1936) *An Analysis of the* De Generatione Animalium *of William Harvey* (Stanford: Stanford University Press).

Middleton, W. E. K. (1971) *The Experimenters: A Study of the Accademia del Cimento* (London: Johns Hopkins University Press).

Mittwoch, U. (2000) 'Three thousand years of questioning sex determination', *Cytogenetics and Cell Genetics*, 91: 186–91.

Moffett, T. (1634) *Insectorum sive Minimorum Animalium Theatrum* (London: Cotes).

Montias, J. M. (1989) *Vermeer and his Milieu: A Web of Social History* (Chichester: Princeton University Press).

Moore, P. (2003) *Blood and Justice* (London: Wiley).

Morange, M. (2001) *The Misunderstood Gene* (Harvard: Harvard University Press).

—— (2003) *La Vie expliquée? 50 ans après la double hélice* (Paris: Odile Jacob).

Muffet (1658) *The Theater of Insects*, in E. Topsel (1658) *The History of Four-footed Beasts and Serpents* (London: Sawbridge, Williams and Johnson).

Mulder, D. (1990) *The Alchemy of Revolution: Gerrard Winstanley's Occultism and Seventeenth-Century English Communism* (New York: Peter Lang).

Needham, J. (1959) *A History of Embryology* (London: Abelard-Schuman).

Nicholl, C. (2004) *Leonardo da Vinci: The Flights of the Mind* (London: Allen Lane).

Nordström, J. (1954–5) 'Swammerdamiana. Excerpts from the travel journal of Olaus Borrichius and two letters from Swammerdam to Thévenot', *Lychnos*, 16: 21–65.

Orme, N. (2001) *Medieval Children* (London: Yale University Press).

Pagel, W. (1958) *Paracelsus* (Basel: Karger).

Palm, L. C, and Snelders, H. A. M. (eds) (1982) *Antoni van Leeuwenhoek 1632–1723* (Amsterdam: Rodopi).

Park, K., and Nye, R. A. (1991) 'Destiny is Anatomy: Review of "*Making Sex: Body and Gender from the Greeks to Freud*" by Thomas Laqueur', *The New Republic*, 18 February 1991, pp. 53–7.

Pasteur, L. (1922) 'Des générations spontanées', in P. Vallery-Radot (ed.) *Oeuvres de Pasteur. Volume 2: Fermentations et générations dites spontanées* (Paris: Masson), pp. 328–46.

Pears, I. (1997) *An Instance of the Fingerpost* (London: Cape).

Petersson, R. T. (1956) *Sir Kenelm Digby: The Ornament of England 1603–1665* (London: Jonathan Cape).

Pineaus, S. (1650) *De Integritatis et Corruptionis Virginum Notis* (Leiden: Moyaert).

Pinto-Correia, C. (1997) *The Ovary of Eve: Egg and Sperm and Preformation* (London: University of Chicago Press).

Pomian, K. (1990) *Collectors and Curiosities: Origins of the Museum* (London: Polity Press).

Portal, A. (1770) *Histoire de l'Anatomie et de la Chirurgie* (Paris: Didot).

Porter, I. H. (1963) 'Thomas Bartholin (1616–80) and Niels Steensen (1638–86) Master and pupil', *Medical History*, 7: 99–125.

Potts, M., and Short, R. (1999) *Ever Since Adam and Eve: The Evolution of Human Sexuality* (Cambridge: Cambridge University Press).

Power, H. (1664) *Experimental Philosophy, In Three-Books* (London: Martin & Allestry).

Price, B. (2002) 'Journeys beyond frontiers: Knowledge, subjectivity and outer space in Margaret Cavendish's *The Blazing World* (1666)', in C. Jowitt and D. Watt (eds) *The Arts of 17th-Century Science* (Aldershot: Ashgate), pp. 127–45.

Principe, L. (1998) *The Aspiring Adept: Robert Boyle and his Alchemical Question* (Princeton: Princeton University Press).

Punnett, R. C. (1928) 'Ovists and animalculists', *American Naturalist*, 62: 481–507.

Pyenson, L., and Sheets-Pyenson, S. (1999) *Servants of Nature* (London: Fontana).

Pyle, A. (2003) *Malebranche* (London: Routledge).

Randolph, T. (1638) 'An Eglogue [sic] to Mr Johnson', in *Poems with The Muses Looking-Glasse and Amyntas* (Oxford: Bowman).

Ray, J. (1671) 'The extract of a letter written by Mr. John Ray to the Publisher from Midleton, 3 July, 1671. Concerning spontaneous generation; also some insects smelling of musk', *Philosophical Transactions*, 6: 2219–20.

—— (1691) *The Wisdom of God Manifested in the Works of the Creation* (London: Smith).

—— (1693) *Three Physico-Theological Discourses* (London: Smith).

—— (1738) *Travels Through The Low-Countries, Germany, Italy and France* (London: Waltude et al.), vol. 1.

Raynalde, T. (1654) *The Birth of Man-kinde; Otherwise Named the Womans Booke* (London: Hood, Roper & Tomlins).

Redi, F. (1664) *Osservazioni intorno alle vipere* (Florence: Stella).

—— *Esperienze intorno alla generazione degl'insetti* (Florence: Stella).

—— (1909) *Experiments on the Generation of Insects* (Chicago: Open Court).

—— (1970) *Expériences sur la génération des insectes et autres écrits de science et de littérature* (Louvain: Bibliothèque de l'Université).

Rey, A. (1930) *De Sylvius à Regnier de Graaf. Quelques considérations sur les idées médicales au XVIIe siècle* (Bordeaux: Imprimerie de l'Académie des Facultés).

Ricci, J. V. (1950) *The Genealogy of Gynaecology* (Philadelphia: Blakiston).

Roberts, B. (2004) 'Drinking like a man: The paradox of excessive drinking for seventeenth-century Dutch youths', *Journal of Family History*, 29: 237–52.

Roberts, B. B., and Groenendijk, L. F. (2004) '"Wearing out a pair of fool's shoes": Sexual advice for youth in Holland's Golden Age', *Journal of the History of Sexuality*, 13: 139–56.

Robinson, H. W., and Adams, W. (eds) (1935) *The Diary of Robert Hooke M.A., M.D., F.R.S.* (London: Taylor & Francis).

Roe, S. A. (1981) *Matter, Life, and Generation: Eighteenth-Century Embryology and the Haller-Wolff Debate* (Cambridge: Cambridge University Press).

—— (2003) 'The Life Sciences', in R. Porter (ed.) *The Cambridge History of Science. Volume 4: Eighteenth-Century Science*, pp. 397–416.

Roger, J. (1963) *Les Sciences de la vie dans la pensée française au XVIIIe siècle* (Paris: Albin Michel).

—— (1981) 'Two scientific discoveries: their genesis and destiny', in M. D. Grmek, R. S. Cohen and G. Cimino (eds) *On Scientific Discovery* (Dordrecht: Reidel), pp. 229–37.

—— (1997) *The Life Sciences in Eighteenth-Century French Thought* (Stanford: Stanford University Press).

Rogers, J. (1996) *The Matter of Revolution: Science, Poetry and Politics in the Age of Milton* (London: Cornell University Press).

Rome, D. R. (1956) 'Nicolas Sténon (1638–1686): Anatomiste, géologue, paléontologiste, cristallographe, vicaire apostolique des régions nordiques', *Revue des questions scientifiques*, 20 October 1956, pp. 517–72.

Rose, S. (1997) *Lifelines: Biology, Freedom, Determinism* (London: Allen Lane).

Rostenberg, L. (1989) *The Library of Robert Hooke: The Scientific Book Trade of Restoration England* (Santa Monica: Modoc).

Rowland, I. D. (2004) 'Athanasius Kircher, Giordano Bruno and the Panspermia of the Infinite Universe', in P. Findlen (ed.) *Athanasius Kircher: The Last Man Who Knew Everything* (London: Routledge), pp. 191–205.

Ruestow, E. G. (1983) 'Images and ideas: Leeuwenhoek's perception of the spermatozoa', *Journal of the History of Biology*, 16: 185–224.

—— (1984) 'Leeuwenhoek and the campaign against spontaneous generation', *Journal of the History of Biology*, 17: 225–48.

—— (1985) 'Piety and the defense of natural order: Swammerdam on generation', in M. J. Osler and P. L. Farber (eds) *Religion, Science and Worldview: Essays in Honor of Richard S. Westfall* (Cambridge: Cambridge University Press), pp. 217–41.

—— (1995) *The Microscope in the Dutch Republic* (Cambridge: Cambridge University Press).

Rupp, J. C. C. (1990) 'Matters of life and death: The social and cultural conditions of the rise of anatomical theatres, with special reference to seventeenth-century Holland', *History of Science*, 28: 263–87.

Sapp, J. (2003) *Genesis: The Evolution of Biology* (Oxford: Oxford University Press).

Sawin, C. T. (1997) 'Regnier de Graaf and the Graafian follicle', *The Endocrinologist*, 7: 415–21.

Schama, S. (1991) *The Embarrassment of Riches* (London: Fontana).

Scherz, G. (ed.) (1969) *Steno: Geological Papers* (Odense: Odense University Press).

Schierbeek, A. (1959) *Measuring the Invisible World* (London: Abelard & Schuman).

—— (1967) *Jan Swammerdam 1637–1680. His Life and Works* (Amsterdam: Swets & Zeitlinger).

Schiller, J., and Théodoridès, J. (1968) 'Sténon et les milieux scientifiques parisiens', *Analecta Medico-Historica*, 3: 155–70.

Schrader, J. (1674) *Observationes et Historiae* (Amsterdam: Wolfgang).

Schulz, W. W., van Andel, P., Sabelis, I., and Mooyaart, E. (1999) 'Magnetic resonance imaging of male and female genitals during coitus and female sexual arousal', *British Medical Journal*, 319: 1596–1600.

Seba, A. (2002) *Cabinet of Natural Curiosities* (London: Taschen).

Seneca (1972) *Naturales quaestiones* (London: Heinemann).

Setchell, B. P. (1974) 'The contributions of Regnier de Graaf to reproductive biology', *European Journal of Obstetrics, Gynecology and Reproductive Biology*, 4: 1–13.

Shadwell, T. (1966) *The Virtuoso* (London: Arnold).

Shapin, S. (1996) *The Scientific Revolution* (London: University of Chicago Press).

—— and Schaffer, S. (1985) *Leviathan and the Air-Pump: Hobbes, Boyle and the Experimental Life* (Chichester: Princeton University Press).

Shepard, A. (1996) '"O seditious Citizen of the Physicall Common-Wealth!": Harvey's royalism and his autopsy of Old Parr', *University of Toronto Quarterly*, 65: 482–505.

Sherley, T. (1672) *A Philosophical Essay* (London: Cademan).

Short, R. V. (1977) 'The discovery of the ovaries', in S. Zuckerman and B. J. Weir (eds) *The Ovary*. Volume 1: *General Aspects* (New York: Academic Press), pp. 1–39.

—— (1978) 'Harvey's conception: "*De generatione animalium*", 1651', in C. J. Dickinson and J. Marks (eds) *Developments in Cardiovascular Medicine* (Lancaster: MTP), pp. 353–63.

—— (2003) 'The magic and mystery of the oocyte: Ex ovo omnia', in A. O. Trounson and R. G. Gosden (eds) *Biology and Pathology of the Oocyte* (Cambridge: Cambridge University Press), pp. 3–10.

Smith, P. H., and Findlen, P. (eds) (2002) *Merchants & Marvels: Commerce, Science and Art in Early Modern Europe* (London: Routledge).

—— (2004) *The Body of the Artisan* (Chicago: Chicago University Press).

St Serfe, T. (1668) *Tarugo's Wiles: Or, the Coffee-House* (London: Herringman).

Stelluti, F. (1630) *Persio: Tradotto in verso sciolto e dichiarato da Francesco Stelluti* (Rome: n.p.).

Steno, N. (1661) *Disputatio anatomica de glandulis oris & nuper observatis inde prodeundibus vasis* (Leiden: n.p.).

—— (1668) *Elementorum Myologiae Specimen* (Florence: Stella).

—— (1950) 'On the passage of yolk into the intestines of the chick. (De vitelli in intestina pulli transitu.) Translated with an introduction and commentary by Margaret Tallmadge May', *Journal of the History of Medicine and Allied Sciences*, 5:119–43.

—— (1965) *Nicolaus Steno's Lecture on the Anatomy of the Brain* (Copenhagen: Nyt Nordisk Forlag).

Stenonis, N. (1675) 'Ova viviparorum spectantes observationes factae', in T. Bartholini (ed.) *Acta Medica & Philosophica Hafniensa* (Hafniae: Haubold & Gödiani), vol. 2, pp. 219–32.

Stephenson, N. (2003) *Quicksilver* (London: Heinemann).

Stevenson, J. (2002) *The Pretender* (London: Cape).

Stone, G. N., and Schönrogge, K. (2003) 'The adaptive significance of insect gall morphology', *Trends in Ecology and Evolution*, 18: 512–21.

Strathern, P. (2005) *The Medici: Godfathers of the Renaissance* (London: Pimlico).

Strick, J. E. (2000) *Sparks of Life: Darwinism and the Victorian Debates over Spontaneous Generation* (London: Harvard University Press).

Swammerdam, J. (1667) *Disputatio Medica Inauguralis, Continens Selectas de Respiratione Propositiones* (Leiden: Elsevir).

—— (1669) *Historia Insectorum Generalis, ofte Algemeene Verhandeling van de Bloedeloose Dierkens* (Utrecht: van Dreunen).

—— (1672) *Miraculum Naturae, sive Uteri Muliebris Fabrica* (Leiden: Severinus Matthei).

—— (1685) *Histoire Generale des Insectes* (Utrecht: Ribbius).

—— (1758) *The Book of Nature* (London: Seyffert) [1978 Facsimile (New York: Arno)].

—— (n.d. 1672?) *Exquisita Demonstratio. Vasorum Spermaticorum, Testium sive Ovarii, Tubarum seu Cornuum, &c.* (n.p.).

—— (n.d.) *Begin. Illustrissimæ . . . Regiæ Societati Londini . . . hoc anatomici sui studii specimen et futuri operis quendam quasi prodromum, . . . dedicat . . . J. S.* (n.p.: n.p.), British Library BM pressmark 748.g.10.(2.).

Swan, C. (2005) *Art, Science and Witchcraft in Early Modern Holland: Jacques de Gheyn II (1565–1629)* (Cambridge: Cambridge University Press).

Tamizey de Larroque, Ph. (ed.) (1883) *Lettres de Jean Chapelain*, vol. 2 (Paris: Impr. Nationale).

Thompson, S. (1957) *Motif-Index of Folk Literature. Volume 5: L–Z* (Copenhagen: Rosenkilde & Bagger).

Tinniswood, A. (2001) *His Invention So Fertile: A Life of Christopher Wren* (London: Cape).

Tomalin, C. (2002) *Samuel Pepys: The Unequalled Self* (London: Penguin).

Topsel, E. (1658) *The History of Four-footed Beasts and Serpents* (London: Sawbridge, Williams and Johnson).

Tribby, J. (1991) 'Cooking (with) Clio and Cleo: Eloquence and experiment in Seventeenth-Century Florence', *Journal of the History of Ideas*, 52: 417–39.

Tucker, H. (2004) *Pregnant Fictions: Childbirth and the Fairy Tale in Early Modern France* (Detroit: Wayne State University Press).

Turner, G. L'E. (2005) 'The impact of Hooke's *Micrographia* and its influence on microscopy', in P. Kent and A. Chapman (eds) *Robert Hooke and the English Renaissance* (Leominster: Gracewing), pp. 124–45.

Underwood, E. A. (1972) 'Franciscus Sylvius and his Iatrochemical School', *Endeavour*, 31: 73–6.

Van der Linden, M. (1997) 'Marx and Engels, Dutch Marxism and the "Model capitalist nation of the seventeenth century"', *Science & Society*, 61: 161–92.

Van der Lugt, M. (2004) *Le Ver, le démon et la vierge: Les théories médiévales de la génération extraordinaire* (Paris: Les Belles Lettres).

Van Helmont, J. B. (1662) *Oriatrike, or, Physick Refined* (London: Loyd).

Van Horne, J. (1668) *Observationum Suarum Circa Partes Generationis in Utroque Sexu Prodromus* (Leiden: Gaasbekios).

—— (1674) *Microcosmus* (Leiden: n.p.).

Van Leeuwenhoek, A. (1941) *The Collected Letters of Antoni van*

Leeuwenhoek, vol. 2 (Amsterdam: Swets & Zeitlinger).

—— (1948) *The Collected Letters of Antoni van Leeuwenhoek*, vol. 3 (Amsterdam: Swets & Zeitlinger).

—— (1957) *The Collected Letters of Antoni van Leeuwenhoek*, vol. 4 (Amsterdam: Swets & Zeitlinger).

—— (1957) *The Collected Letters of Antoni van Leeuwenhoek*, vol. 5 (Amsterdam: Swets & Zeitlinger).

—— (1976) *The Collected Letters of Antoni van Leeuwenhoek*, vol. 9 (Amsterdam: Swets & Zeitlinger).

Van Leewenhoeck (sic), A. (1677) 'Observations, communicated to the Publisher by Mr Antony van Leeuwenhoeck, in a Dutch letter of the 9th of Octob. 1676. Here English'd: Concerning little Animals by him observed in rain- well- sea- and snow-water; as also in water wherein Pepper had lain infused', *Philosophical Transactions*, 12: 821–31.

Van Speybroeck, L., de Waele, D., and Van de Vijver, G. (2002) 'Theories in early embryology: close connections between epigenesis, preformationism, and self-organization', *Annals of the New York Academy of Science*, 981: 7–49.

Von Baer, K. E. (1956) 'On the genesis of the ovum of mammals and of man', *Isis*, 47: 117–53.

Von Baer, K. E. and Sarton, G. (1931) 'The discovery of the mammalian egg and the foundation of modern embryology', *Isis*, 16: 315–77.

Von Haller, A. (1786) *First Lines of Physiology* (Edinburgh: Elliot).

Walton, I. (1653) *The Compleat Angler or the Contemplative Man's Recreation* (London: Marriot).

Wells, W. A. (1949) 'Franciscus Sylvius (François de le Boe), 1614–1672', *The Laryngoscope*, 59: 904–19.

Whitaker, K. (2003) *Mad Madge: Margaret Cavendish, Duchess of Newcastle, Royalist, Writer and Romantic* (London: Vintage).

Whitteridge, G. (1965) 'William Harvey: A royalist and no parliamentarian', *Past & Present*, 30: 104–9.

Wilson, C. (1995) *The Invisible World* (Princeton: Princeton University Press).

Wilson, H. F. and Doner, M. H. (1937) *The Historical Development of Insect Classification* (Chicago: n.p.).

Wolpert, L. (1995) 'Development: Is the egg computable or could we generate an angel or a dinosaur?', in M. P. Murphy and L. O'Neill (eds) *What is Life? The Next Fifty Years: Speculations of the Future of Biology* (Cambridge: Cambridge University Press), pp. 57–66.

Woolley, B. (2004) *The Herbalist: Nicholas Culpeper and the Fight for Medical Freedom* (London: HarperCollins).

Zatylny, C., Marvin, L., Gagnon, J., and Henry, J. (2002) 'Fertilization in *Sepia officinalis*: the first mollusk sperm-attracting peptide', *Biochemical and Biophysical Research Communications*, 296: 1186–93.

Zumthor, P. (1959) *La Vie quotidienne en Hollande au temps de Rembrandt* (Paris: Hachette).

FURTHER READING

The Notes and Bibliography provide references to a large number of primary and secondary sources. However, many readers will undoubtedly wish to pursue the topic by reading recent popular accounts of this period. General background can be found in Lisa Jardine's *Ingenious Pursuits* (1999) and in John Gribbin's *The Fellowship* (2005), both of which are accounts of the work of the Royal Society, and *The Embarrassment of Riches*, Simon Schama's cultural history of the Dutch Golden Age (1991). Clara Pinto-Correia's *The Ovary of Eve* (1995) looks at preformation and covers some of the material presented here, while *The Story of V* by Catherine Blackledge (2003) is a cultural history of the vagina and its representations. The only recent popular work on any of the main characters is *The Seashell on the Mountaintop* by Alan Cutler (2003), which describes the life of Steno, focusing on his contribution to geology. There have been three recent biographies of Hooke, to coincide with the tercentenary of his death – *The Man Who Knew Everything* by Stephen Inwood (2002), *The Curious Life of Robert Hooke* by Lisa Jardine (2003) and *England's Leonardo* by Allan Chapman (2005). Benjamin Woolley's *The Herbalist* (2004) is an account of the life and work of an opponent of Harvey, Nicholas Culpeper. Some recent works of fiction have included some of the characters in this book, including Iain Pears' *An Instance of the Fingerpost* (1997), Jane Stevenson's *The Pretender* (2002) and Neal Stephenson's *Quicksilver* (2003).

PICTURE CREDITS

All pictures are from the author's private collection except:

Page 8: The Royal Collection © 2006 Her Majesty Queen Elizabeth II
Page 22: © The Royal Society
Page 48: © Bibliothèque Centrale, Museum National d'Histoire Naturelle, Paris
Page 57: © Leiden, University Library, ms. Hug. 45
Pages 76 and 79: © Wellcome Institute
Pages 82, 85 and 181: © Bibliothèque Inter-Universitaire de Médicine, Paris
Page 111: © Museum Boijmans Van Beuningen
Page 146: © Göttingen University Library
Page 149: © Leiden, University Library, ms. BPL 126 C1, 125^V–127^R
Page 166: © British Library
Page 194: © Rijksmuseum, Amsterdam
Page 196: © Museum Boerhaave, Leiden

INDEX

Page numbers in **bold** denote illustrations
Page numbers in *italics* denote notes